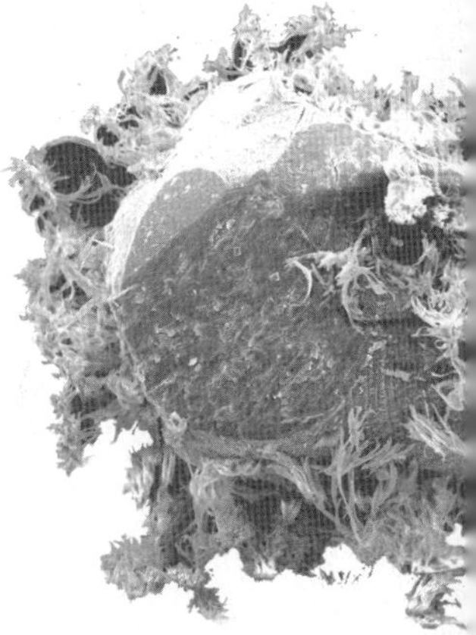

NANOTHERAPEUTICS
Drug Delivery Concepts in Nanoscience

NANOTHERAPEUTICS
Drug Delivery Concepts in Nanoscience

edited by

Alf Lamprecht
University of Franche-Comté, France

PAN STANFORD PUBLISHING

Published by

Pan Stanford Publishing Pte. Ltd.
5 Toh Tuck Link
Singapore 596224

Distributed by

World Scientific Publishing Co. Pte. Ltd.
5 Toh Tuck Link, Singapore 596224
USA office: 27 Warren Street, Suite 401-402, Hackensack, NJ 07601
UK office: 57 Shelton Street, Covent Garden, London WC2H 9HE

British Library Cataloguing-in-Publication Data
A catalogue record for this book is available from the British Library.

NANOTHERAPEUTICS
Drug Delivery Concepts in Nanoscience

ISBN-13 978-981-4241-02-1
ISBN-10 981-4241-02-4

Printed in Singapore by Mainland Press Pte Ltd

In memoriam Armin Lamprecht

Preface

Research and development of innovative drug delivery systems are increasing at a rapid pace throughout the world. This trend will intensify in future as public health expenses demand lower costs and increased efficiency for new therapies. In order to meet this demand, many well-known and efficiently applied drugs will be reformulated in new drug delivery systems that can be value-added for optimized therapeutic activity.

One important aspect in the newly developing field of nanomedicine is the use of nanoparticule drug delivery systems allowing innovative therapeutic approaches. Nanotechnology as a delivery platform offers very promising applications in drug delivery. Due to their small size such drug delivery systems are promising tools in therapeutic approaches such as selective or targeted drug delivery towards a specific tissue or organ, enhanced drug transport across biological barriers (leading to an increased bioavailability of the entrapped drug) or intracellular drug delivery which is interesting in gene and cancer therapy.

The nanotechnological approaches in drug delivery include a large variety of forms, mainly systems based on lipid or polymeric nanoparticles (nanocapsules and nanospheres) microemulsions, liposomes, but also polymeric micelles and cyclodextrins. Potentially different from other scientific communities in the field of drug delivery, nanoparticulates are defined as carrier system with a size below one micron.

On behalf of a great team of nano researchers who have been part of this exciting project, I am pleased to introduce to the scientific community a comprehensive work on Nanotechnology applied in the field of drug delivery, which can be seen as a knowledge base for therapeutic applications of nanotechnologies.

In the past decade, ongoing efforts have been made to develop systems or drug carriers capable of delivering the active molecules specifically to the intended target organ in order to increase the therapeutic efficacy. This approach involves modifying the pharmacokinetic profil of various therapeutic classes of drugs through their incorporation in colloidal nanoparticulate carriers in the submicron size range such as liposomes or nanoparticles. These site-specific delivery systems allow an effective drug concentration to be maintained for a longer interval the target tissue and result in decreased side effects associated with lower plasma concentrations m the peripheral blood. Thus, the principle of drug targeted is to reduce the total amount of drug administered while optimizing its activity. It should be mentioned that the scientific community is still skeptical that such goals could be achieved since huge investments of funds and promising research studies have in many cases resulted in disappointing results and have also been slow in yielding successfully marketed therapeutic nanocarriers. With the recent approval by health authorities of several effective nanosized products containing antifungal or cytotoxic drugs, interest in small drug carriers has been renewed.

A vast number of studies and reviews as well as several books have been devoted to the development, characterization, and potential applications of specific microparticulate- and nanoparticulate delivery systems. No encapsulation process developed to date has been able to produce the full range of capsules desired by potential capsule users. Few attempts have been made to present and discuss in a single book the entire therapeutic range of nanocarriers covered in this book. The general theme and purpose here are to provide the reader with a current and general overview of the existing nanosized delivery systems and to emphasize the various fields of therapeutic applications. The systematic approach used in presenting the first part introducing to the general therapeutic options followed by disease-focused reviewing the existing drug carriers should facilitate the comprehension of this increasingly complex field and clarify the main considerations involved in designing manufacturing, characterizing, and evaluating a specific nanosized-delivery system for a given therapeutic application or purpose.

The first part highlights the exceptional properties of nanoparticles involving their sustained drug release and other physicochemical properties, but especially their ability to trigger drug transport across biological barriers. The general mechanisms of drug delivery, particle translocation, interactions with cells are detailed in this part of the book. Besides, the general strategies of nanoparticulate drug targeting and gene therapy will be elucidated here. The first part of the book starts with a chapter describing the physicochemical aspects of nanocarriers, including particulate systems, liposomes, micellar systems, emulsions, their principal properties, the main excipients necessary for their manufacturing and the basics on their preparation techniques. The authors also address major issues such as the stability of these formulations as well as aspects on the final pharmaceutical form to administer these carriers.

The following chapters deal with the general aspects on drug transport across biological barriers, for the moment one of the most important applications of nanocarriers in the field of therapeutics. Drugs with low permeability properties can significantly enhance their value by their use in a nano-formulation which increases its transport.

Another important aspect is the application of small carriers in the area of drug targeting. This chapter elucidates the potential of nanocarriers in order to allow specific drug delivery to inaccessible disease sites.

The last chapter in this first part is presenting the application of nanodevices in the field of the gene therapy. Although still today most of the gene therapy approaches rely on the use of viral systems, more and more studies deal with the use of non-viral gene delivery due to the advances in the development of biomaterials.

The second part will focus specifically on the therapeutic approaches which are possible by the use of nanocarriers dividing the overall context into chapters dealing with diverse diseases and the relevant therapeutic approaches based on the design of nanoparticulate drug delivery systems.

I am very grateful to all the authors who have shared my enthusiasm and vision by contributing high quality manuscripts, on time, keeping in tune with the original design and theme of this work. You will not be having this book in your hand less their dedication and sacrifice.

Editor
Alf Lamprecht
University of Franche-Comté, France
2007

CONTENTS

Part I

GENERAL ASPECTS OF NANOTHERAPEUTICS

Chapter 1

NANOCARRIERS IN DRUG DELIVERY — DESIGN, MANUFACTURE AND PHYSICOCHEMICAL PROPERTIES

Christoph Schmidt

Biogen Idec International GmbH,
Small Molecule Development International, Zug (Switzerland)

Alf Lamprecht

Laboratory of Pharmaceutical Engineering,
Faculty of Medicine and Pharmacy, University of Franche-Comté,
Besançon (France)

1. Introduction

Colloidal dispersions comprise particles or droplets in the submicron range (< 1 μm Figure 1) in an aqueous suspension or emulsion, respectively. This small size of the inner phase gives such a system unique properties in terms of appearance and application. The particles are too small for sedimentation, they are held in suspension by Brownian motion of the water molecules. They have a large overall surface area and their dispersions provide a high solid content at low viscosity.

The constituents of nanoparticles for biomedical application need to be physiologically compatible (biocompatible), and they need to be biodegradable (disintegrating in physiological environment) to physiologically harmless components or to have the ability to be excreted via kidney or bile.

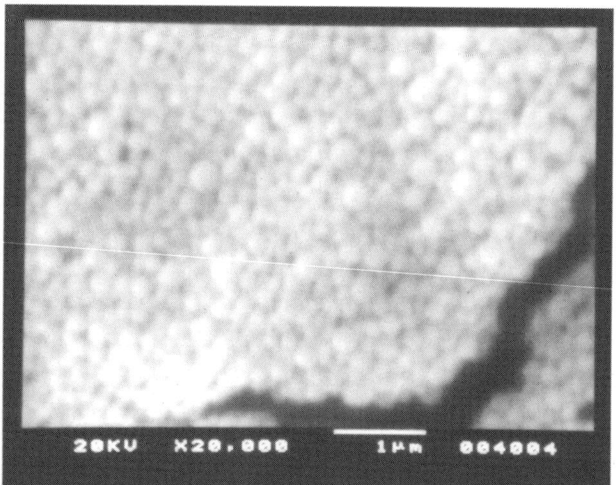

Fig. 1 Scanning electron microscopic image of nanoparticles.

Nanoparticles are carriers for conventional drugs as well as for peptides and proteins, enzymes, vaccines, or antigens. According to the process used for the preparation of nanoparticles, nanospheres or nanocapsules can be obtained. Nanospheres or nanoparticles are homogeneous matrix systems in which the drug is dispersed throughout the particles, whereas nanocapsules are vesicular systems in which the drug is confined to a cavity surrounded by a polymeric membrane (Figure 2).

Fig. 2 Schematic representation of a) nanospheres, drug homogeneously distributed within the matrix, and b) nanocapsules, drug-core surrounded by shell.

Nanocarriers for pharmaceutical use can be of polymeric nature or consist of lipophilic components plus surfactants, i.e., liposomes, niosomes, and solid lipid nanoparticles. Other materials investigated for nanoparticle preparation are albumin [Kreuter, 1994], gelatin [Kreuter, 1994], or calcium alginate [Rajaonarivony *et al.*, 1993]. There are also methods available to size-reduce drug substances into the nanometer range, with the resulting particles being stabilized by surfactants.

Initially, colloidal drug delivery systems have been developed for the intravenous (i.v.) administration of drugs with the goal to improve their therapeutic efficacy through principles like controlled drug release, targeted drug delivery, or prolongation of the circulation time [Kreuter, 1994]. Besides the i.v. route, colloidal particles have also been administered orally, either for systemic uptake or local activity within the gastrointestinal tract [Chen and Langer, 1998; Damgé *et al.*, 1987; Kim *et al.*, 1997; Kreuter, 1991, 1994; Maincent *et al.*, 1986]. In addition, such carrier systems have been developed and tested for almost all routes of administration, for local application on skin and mucosa as well as for systemic use by parenteral application or by inhalation.

This chapter will present an overview over the most prominent examples of colloidal carriers, their different general characteristics, some insights to the preparation of such carriers, as well as describing some details on the physicochemical properties of the diverse systems.

2. Manufacturing of Nanoparticulate Systems

Depending on the nature of the starting material and the intended use of the nanoparticulate system to be prepared, a variety of technologies has been developed and is available for developing and manufacturing colloidal systems. The following overview provides some background information, the most common methods for drugs to be encapsulated and the most important mechanisms of nanoparticle formation from a physicochemical point of view. Various methods have been developed for preparing nanoparticle dispersions as they are established in industries like coating or plastics. However, the application to pharmaceutical systems containing drugs imposes a number of constraints in selecting the materials, for the size of the particles to be prepared, and for the process itself to prevent e.g. drug degradation. Thus, the methods developed in other disciplines have accordingly been adapted to meet these requirements.

2.1. Nanosized Drug Substance

Direct nanosizing of drugs, as reviewed by Merisko-Liversidge *et al.* [2003] provides for delivery of poorly water-soluble drugs to enhance their solubility. Micron-size drug particles are milled in a water-based stabilizer solution for 30-60 minutes to generate nanoparticles with unimodal size distribution. The amount of the suspension stabilizer is critical since too little of it is unable to prevent aggregation of small particles, and too much of it may accelerate particle growth by Ostwald ripening. Since drug dissolution is directly dependant on the surface area, this approach of increasing the specific surface area might be useful for formulation of drugs with a low solubility in aqueous environments.

"Milling" as described above in this context includes processes as conducted in ball- or pearl-mills for a longer time [Liversidge *et al.*, 1992]. Size reduction is obtained by milling pearls made of steel, glass, zircon dioxide, or polymers such as hard polystyrene. Other milling techniques use rotor-stator colloid mills, or jet mills where particles are accelerated and break upon impaction on either another particle or a wall.

High-pressure homogenization, where a suspension of a drug is pressed through a small cavity, is also applied for size reduction of drug particles by shear and impact forces [Müller *et al.*, 2006].

Several other methods have been described in the literature such as the use of supercritical fluid technologies principally leading to particles in the size range of 100 to 500 nm for griseofulvin [Chattopadhyay *et al.*, 2001] or rifampicin [Reverchon *et al.*, 2002]. With supercritical fluids like carbon dioxide, particle formation can be controlled by modifying the pressure which governs solubility of the drugs therein. High pressure generally provides for higher drug solubility, so that upon reduction of the pressure the drug precipitates [Gupta, 2006]. The higher the drop in pressure is, the faster precipitation occurs and in consequence the smaller the resulting particles become. Spraying a solution of drug (alternatively of drug and polymer) in highly compressed supercritical fluid into atmospheric conditions, rapid expansion of this supercritical solution takes place. Instead of supercritical CO_2, organic solvents can be used spray-drying a solution. However, their handling and processing precautions need to be taken into consideration.

Another method to prepare amorphous nanoparticle suspension of poorly water-soluble drugs like Cyclosporine A is evaporative precipitation into aqueous solution. Rapid evaporation of a heated organic solution of the drug is followed by its atomization into aqueous solution. This is leads to a nanoparticle suspension, which can be dried to produce oral dosage forms with low crystallinity and small particle size [Chen *et al.*, 2002].

The aerosol flow reactor method [Eerikainen *et al.*, 2003] involves first dissolving the drug material in question in a suitable solvent, which is then followed by atomizing the solution as fine droplets into a carrier gas. A heated laminar flow reactor tube is used to evaporate the solvent, leaving behind spherical smooth solid drug nanoparticles. The particle diameter increased with increasing reactor temperatures (up to 160°C) due to formation of hollow nanoparticles.

Finally, a high gravity reactive precipitation technique was used to prepare nanoparticles as fine as 10 nm. The feasibility of preparing nanoparticles of organic pharmaceuticals was carried out in rotating packed bed under high gravity. The formation of ultrafine particles was due to intensified micro-mixing of reactants in the rotating bed to enhance nucleation while suppressing crystal growth [Chen *et al.*, 2004].

2.2. Polymeric Nanoparticles

The majority of pharmaceutical nanoparticles are a combination of the active substance with polymeric excipients. Drug-containing nanoparticles can be obtained through the incorporation of a drug substance during or after the preparation of a polymer dispersion. The active components are dissolved, entrapped (in which cases the colloidal particles are often referred to as nanocapsules), or adsorbed to the surface of the nanoparticles. Also, combinations of these arrangements are possible.

For preparation of polymeric nanoparticle dispersions two ways are very common. One is to polymerize the respective monomers in an emulsion or in a micellar system, leading to a system referred to as 'latex'. Such nanoparticlulates are obtained by inducing the reaction of monomers to form the polymeric carrier. The second principal method bases on preparation of nanoparticles using preformed polymers. The latter approach is in general similar to those applied for generating nanosized drug particles, as described above.

2.2.1. Nanoparticles Prepared by Polymerization

Related to the manufacture of latices found in polymer chemistry, methods were adapted from other industrial techniques available for obtaining artificial latexes.

Generally, monomers and suitable catalytic agents are dissolved in an aqueous system comprising either emulsified lipophilic droplets or micelles. At the interface between aqueous and non-aqueous phase or at the surfactant layer of the micelles, respectively, the monomers react with each other leading to oligomers and later polymers. These concentrate in the non-aqueous phase, forming initially soft and semi-solid, subsequently solid particles. Such reactions can occur spontaneously or can be triggered by physical means such as heat or irradiation. The reaction usually continues as long as further monomers are available or, in some cases, as long as reactive groups are present within the polymeric particles. Thus, reactions are terminated by controlling monomer supply or reaction conditions such as temperature, pH, concentration of reactants, or the like. The drug to be entrapped in nanoparticles generated by polymerization is dispersed in either the aqueous phase or in the organic or micellar part, depending on their solubility properties as well as on their susceptibility to interaction with the mono- and oligomers [Speiser, 1998].

However, polymeric nanoparticles prepared by emulsion polymerization may encounter some drawbacks. With the exception of alkylcyanoacrylate, most of the monomers suitable for a micellar polymerization in an aqueous system lead to polymers slowly or not biodegradable. The polymerization process is mainly limited to the vinyl addition reaction, and the molecular weight of the polymeric material cannot be fully controlled. Residues in the polymerization medium (e.g. monomers, oligomers, organic solvents, surfactant, or catalyzing agents) can be toxic and may necessitate further purification of the colloidal material. During the polymerization process, activated monomer molecules also may interact with the drug molecules, potentially leading to their inactivation or modification [Grangier *et al.*, 1991].

Nonetheless, emulsion polymerization is a very popular approach used to synthesize polymer colloids with a matrix structure. This process for polymerization of polyalkylcyanoacrylates was introduced by Couvreur *et al.* [1979] to design biodegradable nanoparticles for the delivery of drugs with various physico-chemical properties.

Methods based on interfacial polymerization have been developed to prepare nanocapsules consisting of a liquid core surrounded by a thin polymer envelope [Al Khoury-Fallouh *et al.*, 1986]. The reactions are performed either in water-in-oil or in oil-in-water emulsion systems, or in microemulsions, leading to the production of water- or oil-containing nanocapsules, respectively. Oil-containing nanocapsules are obtained by the polymerization of alkylcyanoacrylates at the oil/water interface of a very fine oil-in-water emulsion [Al Khoury-Fallouh *et al.*, 1986]. Water-containing nanocapsules may be obtained by the interfacial polymerization of alkylcyanoacrylate in water-in-oil microemulsions. In these systems, water-swollen micelles of surfactants of small and uniform size are dispersed in an organic phase. The monomer is added to the microemulsion and polymerizes at the surface of the micelles. The polymer forms locally at the water-oil interface and precipitates to produce the nanocapsule shell [Gasco and Trotta, 1986].

2.2.2. Nanoparticles Prepared by Preformed Polymers

Beside the already mentioned toxicological aspects, not many polymeric materials are capable of being prepared by emulsion polymerization, examples are polyurethane, epoxyethers, polyester, and others, including semi-synthetic polymers such as cellulose derivatives. Such materials remain unavailable for aqueous dispersion. In order to overcome some of these limitations, nanoparticle preparation methods using various preformed macromolecular materials have been developed. The physicochemical and biological properties of the polymers formed by conventional polymeric synthesis pathways can be well controlled. Dispersion formation from such materials leads to so called pseudo-latices (= artificial latices) [Kreuter, 1994]. For drug delivery purposes, the polymeric material needs to meet physicochemical and biological needs to have its physicochemical and biological properties adapted and optimized for their specific application. Most of the synthetic latices prepared for industrial applications did not meet these requirements, and various approaches were developed in order to obtain polymeric nanoparticles fulfilling the pharmaceutical criteria.

The corresponding manufacturing techniques to obtain colloidal systems generally dissolve the preformed polymer in an organic water-miscible or -immiscible solvent or in a supercritical fluid, i.e. in a gas held under high pressure. This solution is emulsified into water, and the solvent is evaporated or controlled desolvation is applied. To obtain particles in the nanometer range, it is essential to decrease the droplets of the emulsion to the desired size. High shear equipment is employed, using e.g., high pressure homogenization [Gurny *et al.*, 1981], sonication [Krause *et al.*, 1985], or microfluidization [Bodmeier and Chen, 1990].

The methods of preparing pharmaceutical nanoparticles aim at processing different water-insoluble polymeric materials, they are rarely specific to certain polymers. Almost all of the industrial techniques for obtaining artificial latexes rely on one of the following processes:

2.2.2.1 Emulsion-Solvent Evaporation

Polymer and drug are dissolved in a suitable volatile solvent which is immiscible with water. This solution is emulsified in an aqueous solution containing stabilizer (mostly surfactants) by conventional emulsification techniques. Droplet size can be further reduced by using a high-energy source. Continuous emulsification under mixing prevents coalescence of organic droplets and allows the spontaneous evaporation of the solvent at room temperature and the formation of the colloidal particles. Following evaporation of organic phase under reduced pressure or vacuum produces a fine aqueous dispersion of nanoparticles. These can be collected by centrifugation, washed to remove residual stabilizer and can be freeze dried for storage [Quintanar-Guarrero *et al.*, 1998; Jaiswal *et al.*, 2004; Song *et al.*, 1997]. As this approach is limited to certain, mainly water-insoluble drugs, a variation of the first method has been developed for the encapsulation of more hydrophilic drugs. The so-called double-emulsion-technique is thus very interesting for the entrapment of peptides, proteins and nucleic acid sequences. Here, water-in-oil-in-water (w/o/w) emulsions are used, incorporating hydrophilic drugs in an inner aqueous phase [Vandervoort *et al.*, 2002]. The polymer is dissolved in the organic phase and a first mixing step forms a water-in-oil emulsion which is thereafter emulsified in a second, outer aqueous phase. Upon evaporation of the organic phase the polymer precipitates on the surface of the inner aqueous droplets, thereby entrapping the drug dissolved therein.

This technique was principally applied to the preparation of particles from water insoluble polymers. Since in recent years some interesting biopharmaceutical properties were observed with highly hydrophilic polymers, adapted preparation methods were described. In contrast to the prior methods, hydrophilic polymers are dissolved in an aqueous inner phase and emulsified in a non-miscible apolar liquid, and either the inner aqueous phase is eliminated under reduced pressure or the polymer was solidified by a cross-linking reaction [Mitra *et al.*, 2001]. Solidified particles are obtained after washing and drying steps.

Control of droplet size and size distribution of the emulsion are very important factors in the preparation of nanoparticles by these processes. This warrants reproducibility and quality control especially if the process has to be scaled-up. High pressure homogenizers are capable of rapidly and reproducibly forming emulsions in the required (nano-) size range. The equipment finds applicability in other methods of preparation and is available from many suppliers, suited for different scales of production. It has been explored by many researchers for producing nanoparticles in a narrow size range.

2.2.2.2 Solvent-displacement, -diffusion, or Nanoprecipitation

A solution of polymer, drug and lipophilic stabilizer (surfactant) in a semi-polar solvent miscible with water is injected into an aqueous solution (being a non-solvent or antisolvent for drug and polymer) containing another stabilizer under moderate stirring. Nanoparticles are formed instantaneously by rapid solvent diffusion and the organic solvent is removed under reduced pressure [Kumar *et al.*, 2004]. The velocity of solvent removal and thus nuclei formation is the key to obtain particles in the nanometer range instead of larger lumps or agglomerates [Gupta, 2006 a]. As an alternative to liquid organic or aqueous solvents, supercritical fluids can be applied [Gupta, 2006 b].

Fessi *et al.* [1986] proposed a simple and mild method yielding nanoscale and monodisperse polymeric particles without the use of any preliminary emulsification. Both, solvent and nonsolvent must have low viscosity and high mixing capacity in all proportions, like e.g. acetone and water. Another delicate parameter is the composition of the solvent/polymer/water mixture limiting the feasibility of nanoparticle formation. The only complementary operation following the mixing of the two phases is to remove the volatile solvent by evaporation under reduced pressure.

One of the most interesting and practical aspect of this methods is its capacity to be scaled up from laboratory to industrial amounts, since they can be run with conventional equipment.

2.2.2.3 Salting-out

Although a less common method of preparation, by adding a solution of polymer and drug in a water miscible solvent to an aqueous solution containing a salting-out agent and a stabilizer under stirring, small droplets can be obtained. The salting-out agent reduces the solubility of the drug and polymer in water. Dilution of the resulting o/w emulsion with water forces diffusion of organic solvent into the aqueous phase. The remaining polymer together with the drug produces particles in the nano-size range [Allémann *et al.*, 1993 b]. The resulting dispersion often requires a purification step to remove the salting-out agent [Ibrahim *et al.*, 1992; Allémann *et al.*, 1992]

There are many other nanoparticle preparation methods and the few techniques shown above can only give an idea about the most common ones. A more in-depth insight into the particle preparation technologies can be found in the respective literature.

2.2.3. Materials for Preparing Polymeric Nanoparticles

Nanoparticle formulation chemistries have produced a wide spectrum of polymer structures, which are suitable for encapsulation, delivery, and controlled release of both, low molecular pharmaceuticals and biotechnological drugs. Of primary concern are considerations of toxicity, irritancy and allergenicity, and the need for a biodegradable or soluble material.

Polymers used for parenteral delivery have to be biodegradable and are mostly based on polyacrylates (e.g., polycyano-acrylates) [Kreuter, 1983; Couvreur and Vauthier, 1991] or polyesters (e.g., polylactides) [Allémann *et al.*, 1993 a; Brannon-Peppas, 1995].

A number of different polymers have been evaluated for the development of oral vaccines, including naturally occurring polymers (e.g., starch, alginates and gelatin) and synthetic polymers (e.g., polylactide-co-glycolides (PLGA), polyanhydrides, polycyanoacrylates, and phthalates).

Natural Polymers

The advantages of using natural polymers include their low cost, biocompatibility, and aqueous solubility. However, the natural polymers may also be limited in their use due to the presence of extraneous contaminants, variability from batch to batch, and usually low hydrophobicity to entrap lipophilic drug substances.

Natural polymers offer the advantage of established history of safety and use and a high compatibility with both, the human body as well as drugs and other formulation components. Mostly they are water-soluble, but can be transformed into nanoparticles by means of denaturation, leading to cross-linking and thus reduced water solubility. In case of charged groups being present in the material, the use of oppositely charged counter-ions also leads to formation of particles by electrostatic neutralization. Often this is also referred to as coacervation.

Albumin, being established as a protein substitute for human use bears the advantage of complete compatibility even at high amounts, and it also provides surface active properties making it well suitable for stabilization of polymeric nanoparticle [Bazile *et al.*, 1992]. Similarly it could be shown to stabilize manufacturing a respective preparation for paclitaxel [Desai *et al.*, 1999]. Albumin can form layers on drug nanoparticles, which are stabilized by denaturation of the protein. This denaturation can be introduced by cross-linking agents such as aldehydes, or it can be initiated by shear forces as they are applied during processes like emulsion-evaporation (see above).

Gelatin, also widely used in pharmaceutical preparations, can be similarly to albumin processed to reveal proteinic nanoparticles. It can be the major constituent of nanoparticles, embedding the drug, or it can be deposited on the surface of nanoparticles consisting of drug, or drug and polymer, respectively. The different types of gelatin thereby allow for a variety of possibilities to find the best suitable one for the particles and/or manufacturing process in question.

Chitosan ((1→4)-2-amino-2-deoxy-β-D-glucan) is a deacetylated chitin that is of great interest as a functional material of high potential in various areas including the biomedical field. Artursson *et al.* [1994] reported that chitosan can increase the paracellular permeability of intestinal epithelia which attributed to chitosan polymers the property of transmucosal absorption enhancement. Because of low production costs, biocompatibility, and very low toxicity, chitosan is a very interesting excipient for vaccine delivery research. An important advantage of chitosan nano- and microparticles is that, often, the use of organic solvents, which may alter the immunogenicity of antigens, is avoided during preparation and loading [van der Lubben *et al.*, 2001].

Synthetic Polymers

Biodegradable polymers have been extensively used in prolonged parenteral drug delivery as they have the advantage of not requiring surgical removal after they serve their intended purpose.

Thus, most nanoparticles are based on synthetic or semi-synthetic polymers, due to their reproducible manufacture and good stability. They can be synthesized in a wide range of chain length as well as with side chain type and number. By this tailoring towards the desired degradation rates, molecular weights, and co-polymer compositions, the performance of the polymer can be adapted to the intended application.

In addition, by selecting suitable chemical composition and molecular structure, polymeric nanoparticles can be designed to provide properties such as thermo- or pH-sensitivity, or sensitivity to other environmental conditions. This allows targeting drug release to sites within the body having specific conditions, to which the nanoparticles respond [Qiu and Bae, 2006].

Nevertheless, synthetic polymers may be less advantageous due to their limited solubility in physiologically compatible liquids. They are often soluble only in organic solvents and, depending on their structure most synthetic polymers are highly lipophilic and require additional excipients, i.e. surfactants, to form stable nanoparticle dispersion [Singh and O'Hagan, 1998].

Polyglycolic acid, PGA, polylactic acid, PLA, and especially their copolymers PLGA of different ratio and molecular weight are the most commonly used family of biodegradable polymers [Edlund *et al.*, 2003]. The PLGA copolymer is degraded in body by hydrolytic cleavage of ester linkage into lactic acid and glycolic acid at a very slow rate. The acids are easily metabolized in the body via Krebs' cycle and are eliminated as carbon dioxide and water [Panyam *et al.*, 2003].

Polylactic acid (PLA) was among the first polymers being used for biodegradable implants. Polymer chains are cleaved by hydrolysis, leading to water soluble and physiological lactic acid as metabolite.

Polylactide coglycolide (PLGA) is widely used as suitable matrix for drug delivery nanoparticles due to its ease of preparation, commercial availability at reasonable cost, versatility, biocompatibility, and hydrolytic degradation into absorbable and physiologically compatible products. The popularity of PLGA is further enhanced by the fact that FDA as well as European regulatory authorities have approved PLGA for a number of clinical applications [Edlund *et al.*, 2003].

Poly(ε-caprolactone), PCL, is also recognized as a biodegradable and nontoxic material. Because PCL, especially from polymers with high molecular weight, hydrolyses more slowly compared to PLA and PLGA, it is more suitable for long-term drug delivery. The degradation products are neutral in nature and do not interfere with the pH-balance in human tissue. Another valuable property of PCL is its remarkable compatibility with numerous other polymers, allowing for tailoring the properties of the resulting formulation by adding other constituents.

The polyalkylcyanoacrylate nanoparticle family comprises nanospheres, oil- and water-containing nanocapsules and core-shell nanospheres. Their properties are mainly controlled by the side-chains introduced. In general, the longer the alkyl side chains, the longer the half life of particle degradation in vivo. Another influencing parameter of the degradation kinetic are the properties of the alkyl group that modify the hydrophobicity in the order polybutylcyanoacrylate > polyethylcyanoacrylate > polymethylcyanoacrylate and can have consequently an impact on the drug release behavior [Kreuter, 1983].

3. Lipid Based Colloidal Systems

Given the considerations concerning toxicology of polymers and their degradation products, more physiologic components with suitable solubility for lipophilic drugs can be found in the field of pharmaceutical lipids. These comprise e.g. triglycerides, being physiological components, and are usually well biodegradable and thus exhibit low toxicity [Müller, 1998 b; Heurtault *et al.*, 2003]. They resemble oil-in-water emulsions, but with the internal phase being small in size and in many cases of solid consistency. Another lipid based colloidal system are liposomes, vesicular structures akin to cell membranes.

3.1 Solid Lipid Nanoparticles

Colloidal particles consisting of solid triglycerides or other lipid substances were first produced by dispersion of molten lipids by means of high-shear or ultrasound [Speiser, 1990]. Similarly, preparation of a microemulsion (see below) at higher temperatures can lead to solidification of the lipid phase upon cooling and thus to a dispersion of colloidal lipid particles.

A method also applicable on larger scale is high pressure homogenization [Müller and Lucks, 1996; Müller, 1998 b]. The process is run either at elevated temperatures with molten lipids in aqueous dispersion or at lower process temperatures where solid lipids are broken down into nanosized particles when pumped through the small gap in the homogenizer.

Solid lipid nanoparticles can alternatively be prepared by rapidly injecting a solution of solid lipids in a water miscible solvent mixture into water to get particles of 80-300 nm [Schubert and Müller-Goymann, 2003; Arica Yegin *et al.*, 2006]. Another group used a particle engineering process of spray-freezing into liquid to generate a rapid dissolving high potency danazol powders of 100 nm [Hu *et al.*, 2004].

Besides the main component, a solid lipid material serving to dissolve or disperse the drug incorporated, SLN often require surfactants for their stabilization, i.e. to prevent aggregation and to enable a nanosized dispersion being generated during processing. Also, these surfactants lead to more round particles, whereas plain lipids generally form cubic crystal-like particles [Müller, 1998 b].

A relatively recent development are the lipid nanocapsules (LNC) prepared by a phase inversion method [Heurtault *et al.*, 2002; Lamprecht *et al.*, 2002]. This is a solvent-free preparation method leading to small capsules in the size range of 20 to 100 nm.

3.2. Liposomes

Vesicular carriers comprising a hydrophilic core surrounded by one or more lipid bilayer membranes were for the first time described by Bangham *et al.* 1965 (Figure 3). Initially, they were used as models for physiological membranes [Bangham, 1968] before being considered for as drug carriers [Gregoriadis, 1974; Papahadjopoulos and Vail, 1978].

The bilayer consists typically of phospholipids (lecithins), cholesterol, and glycolipids, having a thickness of about 5 nm. Liposomes can be produced in sizes from below 50 nm up to several μm depending on the composition and the manufacturing process [Schubert, 1998]. They can carry hydrophilic drugs within their core as well as lipophilic substances being dissolved or dispersed in the membrane.

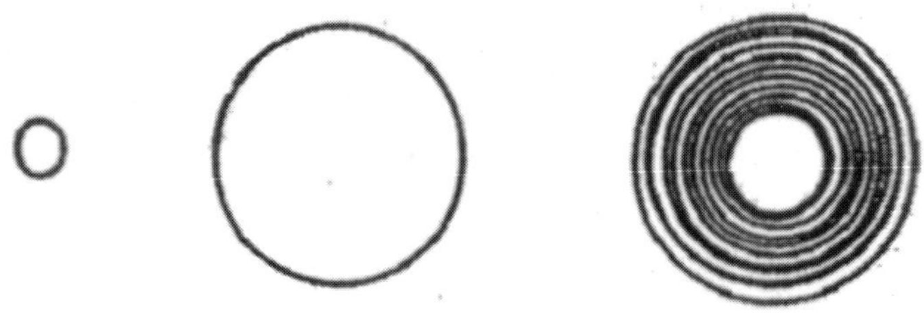

Fig. 3 Schematic overview over various bilayer arrangements in liposomes. SUV (left), LUV (middle), and MLV (right).

Small unilamellar vesicles (SUV) have the core encapsuled by one layer and have a size of generally up to 50 nm. The corresponding unfavorable energy status associated with the high curvature of the bilayer [Thompson *et al.*, 1974] is to the most part compensated by the outer monolayer bearing more lipid molecules than the inner layer [de Kruijff *et al.*, 1975]. Energetically preferred are large unilamellar vesicles, LUV, having a monolayer without much tension surrounding a larger core. In case of several concentric monolayers with aqueous interstitial volumes surrounding a likewise aqueous core, one refers to these structures as multilamellar vesicles (MLV). They are especially apposite for sustaining the release of hydrophilic drugs, which have to penetrate several lipophilic layers [Schubert, 1998].

There is a wide variety of manufacturing methods described, of which those using mechanical means to produce the vesicles are preferred for industrial use due to their ability to be well controlled and reproducible. Examples are ultrasound [Huang, 1969], high-pressure homogenization [Barenholz *et al.*, 1979], or extrusion through a membrane filter [Olson *et al.*, 1979, Mayer *et al.*, 1986]. The energy input leads to dispersion of the lipids, which reassemble to the described membrane-like structures to reduce interfaces.

Spontaneous formation of colloids, i.e., the preparation of a dispersion without using high-shear equipment to reduce size, was also successfully applied for liposomes. Lipids were dissolved in ethanol or other solvents and injected into a water phase. This so called solvent injection method [Batzri and Korn, 1973] revealed unilamellar vesicles, with the size being a function of the applied solvent. When the bilayer-forming lipids were dissolved in ethanol, small vesicles were obtained while water-immiscible solvents led to large liposomes.

Liposomes are used for solubility enhancement in parenteral formulations, to allow for a higher amount of drug to be administered and to circulate in the bloodstream. Also, they are widely used in dermatological preparation as well as in cosmetics due to their ability to penetrate into deeper skin levels. For oral use liposomes are in many cases not suitable, their susceptibility to stomach-pH and instestinal enzymes renders do not allow to make use of their properties for oral medication.

Similar in structure, configuration and also in preparation are niosomes, which comprise synthetic, non-ionic lipids instead of phospholipids. These constituents by their nature exhibit higher chemical stability [Schubert, 1998] while otherwise maintaining the general properties of liposomes.

4. Microemulsions

Although their structure is assumed not to be of particulate nature, microemulsions are also considered as a colloidal system, with unique properties making them useful for drug delivery. They are close to spontaneously formed nanoparticles in terms of preparation as well as to micellar systems in terms of properties. The name "microemulsion" does not refer to them comprising an inner phase in the micrometer range as it might suggest. It is instead most likely derived from their composition being similar to conventional emulsions, albeit they do have distinctly different properties.

Microemulsions comprise two immiscible liquids and at least one emulsifying agent, mostly applied together with a cosurfactant. Due to the ratio of oil and water, they cannot be considered micellar solutions [Pouton, 1997]. Macroscopically, they appear as clear, one-phase isotropic systems. The dispersed phase consists of very small particles (5 - 140 nm [Attwood, 1994]) and its properties resemble those of a bulk phase rather than an inner emulsion phase. The enormous reduction in interfacial tension, enabling the large interface, is provided by a high amount of surfactant and cosurfactant. This low interfacial tension also supports the spontaneous formation of such systems, not requiring any energy input. Consequently, as a thermodynamically stable system, microemulsions do not exhibit stability problems such as phase separation or increase in particle size.

Systems without an aqueous phase, i.e., surfactant and cosurfactant dissolved or dispersed in oil, are known as self-emulsifying or self-micro-emulsifying drug delivery systems (SEDDS or SMEDDS). They form microemulsions upon exposure to aqueous media, as in the gastro-intestinal tract [Constantinides, 1995]. Depending on the composition, with an excess of the aqueous phase, no transparent microemulsion is formed, but an opaque conventional emulsion - but without energy input.

Formulation strategies for microemulsions are reviewed by e.g. Constantinides [1995] and Pouton [1997]. Medium-chain triglycerides are good candidates to start with. They are stable, recognized as safe by regulatory authorities, and improve drug absorption. For many drugs, they enable the development of alcohol-free formulations, beneficial in terms of toxicity and stability against evaporation [Constantinides *et al.*, 1994]. For stabilization, nonionic surfactants are preferred due to their lower toxicity [Constantinides, 1995; Hochman and Artursson, 1994].

However, the use of microemulsions is associated with some drawbacks, limiting their use to the application of problematic drugs rather than being a universal tool [Attwood, 1994]: The required high amount of surfactant decreases tolerability after application; stability problems might occur with the use of ethanol (volatile) and oils (rancidity); a final dosage form like a capsule is likely to suffer from incompatibilities between capsule shell and oils and/or surfactants; possible incomplete solubilization of drug leads to drug precipitation; only a limited drug load (10 - 15%) can be achieved, precluding drugs with higher doses from being incorporated into such a system.

5. Polymeric Micelles

The capacity of block copolymer micelles to increase the solubility of hydrophobic molecules stems from their unique structural composition, which is characterized by a hydrophobic core sterically stabilized by a hydrophilic corona (Figure 4). The former serves as a reservoir in which the drug molecules can be incorporated by means of chemical, physical or electrostatic interactions, depending on their physicochemical properties [Jones *et al.*, 1999].

Beyond solubilizing hydrophobic drugs, block copolymer micelles can also target their payload to specific tissues through either passive or active means. Prolonged in vivo circulation times and adequate retention of the drug within the carrier are prerequisites to successful drug targeting. Long circulation times ensue from the steric hindrance awarded by the presence of a hydrophilic shell and the small size (10–100 nm) of polymeric micelles. Indeed, micelles are sufficiently large to avoid renal excretion (> 50 kDa), yet small enough (< 200 nm) to bypass filtration by inter-endothelial cell slits in the spleen. Drug retention, in turn, is dependent on micelle stability and polymer–drug interactions. Many approaches are being employed to enhance the physical stability of

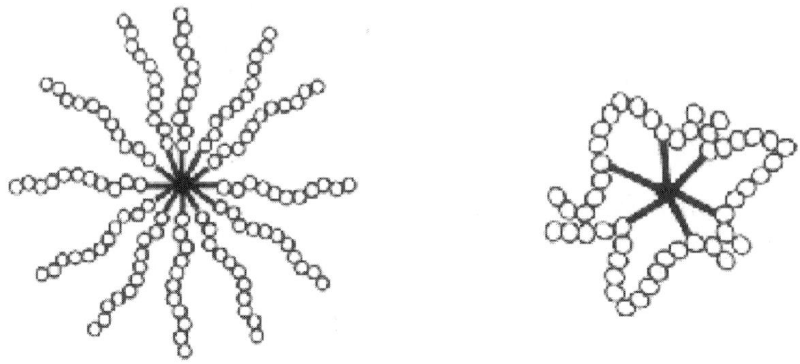

Fig. 4 Schematic representation of polymeric micelles [Jones *et al.*, 1999].

the carrier, improve its resistance towards dissociation upon entering the bloodstream, and tailor its properties to better suit those of the incorporated drug.

The self-assembly of amphiphilic block copolymers in water is based on non-polar and hydrophobic interactions between the lipophilic, core-forming polymer chains. Most amphiphilic copolymers employed for drug delivery purposes contain either a polyester or a poly(amino acid)-derivative as the hydrophobic segment. Polyethers constitute another class of polymers that can be employed to prepare amphiphilic micelles. Most of the polyethers of pharmaceutical interest belong to the poloxamer family, i.e. block-copolymers of polypropylene glycol and polyethylene glycol. Depending on the physicochemical properties of the block copolymer, two main classes of drug-loading procedures can be applied. The first class, direct dissolution, involves dissolving the copolymer along with the drug in an aqueous solvent. This procedure is mostly employed for moderately hydrophobic copolymers, and may require heating of the aqueous solution to bring about micellization via the dehydration of the core-forming segments.

The second category of drug-loading procedures applies to amphiphilic copolymers which are not readily soluble in water and for which an organic solvent common to both the copolymer and the drug (such as dimethylsulfoxide, *N,N*-dimethylformamide, acetonitrile, tetrahydrofuran, acetone or dimethylacetamide) is needed. The mechanism by which micelle formation is induced depends on the solvent-removal procedure. For water-miscible organic solvents, the copolymer mixture can be dialyzed against water, whereby slow removal of the organic phase triggers micellization. Alternatively, the solution-casting method entails evaporation of the organic phase to yield a polymeric film where polymer–drug interactions are favored. Rehydration of the film with a heated aqueous solvent produces drug-loaded micelles. Physical entrapment of a hydrophobic drug may be further achieved through an oil-in-water (O/W) emulsion process which involves the use of a non-water-miscible organic solvent (dichloromethane, ethyl acetate). The above-mentioned techniques all require sterilization and freeze-drying steps to produce injectable formulations with an adequate shelf-life.

Process parameters such as the nature and proportion of the organic phase, as well as the latter's affinity for the core-forming segment, can affect the preparation of drug-loaded polymeric micelles and alter the properties of the end product. In addition, the incorporation method itself can modulate the attributes of the yielded micelles.

6. Factors Affecting Certain Carrier Properties

To achieve the desired or required properties for nanoparticles to be prepared, an understanding of some general principles of manufacture and composition is beneficial. This allows focused formulation of colloidal drug preparations. The following section provides an overview over the existing data in the field.

6.1. Drug Loading

Among the influencing factors on the extent of drug loading are method of preparation, additives (e.g. stabilizers, bioadhesives including mucoadhesives, solvent), nature of drug and polymer, their respective solubilites, and pH. Formulation variables can be modulated to increase the drug loading in nanoparticles [Govender *et al.*, 1999]. Depending on both the preparation process and the physicochemical properties of both the drug molecule and the carrier, the drug entrapment can be either by inclusion within the carrier and/or by surface adsorption onto this carrier.

Any kind of preparation process, polymerization of monomers or dispersion of preformed polymer, entrapment within non-porous NP requires the solubility of the drug molecule in the macromolecular material, whereas porous nanoparticles may entrap the drug molecule by adsorption either onto the surface or within the macromolecular network. Entrapment within the core of nanocapsules implies the solubility of the drug molecule in the oily phase used during preparation. It should be mentioned that the drug to polymer ratio can be as large as 500:1 in nanocapsules (inner core made of the drug itself) when this ratio is usually under 10% in nanospheres. Electrical charges on both, the drug molecule and the carrier may influence the loading capacity. The adsorption of drugs onto nanoparticles can be described following the Langmuir-type or the constant partitioning-type isotherms. In fact, nanoparticles generally entrap drug molecules according to a Langmuir adsorption mechanism owing to their large specific surface area.

As a promising approach, Li *et al.* [2004] have prepared porous hollow silica nanoparticles, which can be used to incorporate drugs in higher doses, allowing for dosage forms with smaller volume at a given dose.

6.2. Drug Release

Drug release from colloidal carriers is dependent on both the type of carrier and the preparation method applied. As the specific surface area of a nanodispersion is very large, the release rate may be more rapid than from larger structures such as microcapsules. Thus, colloidal carriers are usually not able to act as long-term sustained-release delivery systems.

Depending on the polymer selected and the method of manufacture, drug release from nanoparticles usually follows the following mechanisms: desorption from surface, diffusion through matrix or wall, or erosion of the matrix. In nearly all cases a combination of these phenomena occurs.

Release from nanoparticles may be different according to the drug-entrapment mechanism involved. When the drug is adsorbed on the particle surface, the release mechanism can be described as a partitioning process. When the drug is entrapped within the matrix, diffusion plus bioerosion are involved in the release mechanism, whereas diffusion will be the main mechanism if the carrier is not biodegradable.

In many cases, drug release from nanoparticles was observed to be biphasic - an initial burst is followed by a rather slow (thus controlled) release. Although this pattern seems universal, Rosca *et al.* [2004] have offered an explanation for this phenomenon for nanoparticles prepared by emulsification solvent evaporation method: With single emulsions, the solvent elimination concentrates the incorporated substance towards the surface and for multiple emulsions, it makes holes in the polymeric walls near the surface, resulting in the initial burst release. The rest of the incorporated drug is released under the dual influence of diffusion within the matrix and polymer degradation.

The active drug can also be bound chemically to a suitable carrier polymer. In such instances, drug release is governed by the cleavage of these chemical bonds, e.g. by hydrolysis or by enzymatically catalyzed reactions. Similar to pro-drug concept, the active moiety is generated after application of the medication.

Special release mechanisms can be procured by selecting polymers having distinct properties with regard to chemical composition and molecular structure. Proper selection of these features like thermo sensitivity or pH-dependency allow to tailor drug release to respond to environmental effects.

7. Stability and Storage

A pharmaceutical formulation faces various stability challenges during preparation, storage and even after administration, before the drug included can be delivered to the targeted site of action.

Depending on its chemistry and morphology, a polymer will absorb some water on storage in a humid atmosphere. Absorbed moisture can initiate degradation and a change in physicochemical properties, which can in turn affect the performance in vivo. Storage conditions may thus be critical to the shelf life of a polymeric nanoparticle formulation. The presence of oligomers, residual monomer, or remaining polymerization catalysts or solvents may impair the storage stability, catalyzing moisture absorption or degradation. The incorporation of drug may also affect the storage stability of a polymer matrix. The relative strength of water polymer bonds and the degree of crystallization of polymer matrix are other important factors. To maintain absolute physicochemical integrity of degradable polymeric drug delivery device, storage in an inert atmosphere is recommended [Edlund *et al.*, 2002].

Commercialization of liquid nanoparticulate systems has not taken up partly due to the problems in maintaining stability of suspensions for an acceptable shelf life [Saez *et al.*, 2002]. The colloidal suspension, in general, does not tend to separate just after preparation because submicronic particles sediment very slowly and the aggregation effect is counteracted by mixing tendencies of diffusion and convection. However, after several months of storage, aggregation can occur. Additionally, microbiological growth, hydrolysis of the polymer, drug leakage and/or other component degradation in aqueous environment is possible. Freeze-drying is a good method to dry nanoparticles in order to increase the stability of these colloidal systems. However, due to their vesicular nature, especially nanocapsules are not easily lyophilized, as they tend to collapse, releasing the core content.

Stability of polybutylcyanoacrylate nano-suspensions was examined by measuring particle sizes and size distributions over a period of 2 months in hydrochloric acid, phosphate buffered saline (PBS) and human blood serum. When stored in acidic medium, nanoparticles were found to be stable for at least two months while those stored in PBS agglomerated and showed increase in their polydispersity index. When added to human blood serum, nanoparticles were found not to agglomerate, remaining stable in size for at least five days. Thus instead of lyophilization, which potentially poses problems with reconstitution, acidic storage can ensure stability in certain cases [Schroeder *et al.*, 1998].

Freeze-dried poly(methylidene malonate) (PMM) nanoparticles were evaluated for their 12-months stability under various storage conditions with respect to temperature and exposure to light. Alterations in nanoparticles kept at 40°C were explained on the basis of degradation of the polymer side chains and generation of carboxyl moieties. Lyophilized PMM colloidal nanoparticles stored at room temperature or below, either in darkness or in daylight were claimed to have a satisfactory shelf-life of one year [Breton *et al.*, 1998].

As an example for lipid based nanoparticles, stability of a surfactant stabilized SLN formulation was investigated as a function of storage temperature, exposure to light, and type of glass container (untreated and siliconized glass vials). Exposure to energy (temperature, light) led to particle growth and subsequent gelation in the system. The type of glass did not have much effect while siliconization of the vials almost eliminated particle growth. By optimization of the storage conditions, stability of over three years was claimed [Freitas and Müller, 1998].

8. Nanoparticle-containing Dosage Forms

For parenteral applications, colloidal dispersions are commonly used as such or converted to the dry state by means of lyophilization [Allémann *et al.*, 1993 a] and redispersed prior to administration.

For oral delivery into the human body nanoparticles can be also administered as their aqueous dispersion as the final dosage form. This is a way of delivery without further processing after nanoparticle formation. However, poor stability of the drug or polymer in an aqueous environment or poor taste of the drug may require the incorporation of the colloidal particles into solid dosage forms, i.e. into capsules and tablets.

Colloidal particles can be incorporated into solid dosage forms either in solid or liquid form. The dispersion of the colloidal particles can be dried (i.e., spray- or freeze-dried), if needed together with suitable excipients, followed by filling of the dried powder into capsules or compressing it into tablets [Allémann *et al.*, 1993 a]. Suitable conventional excipients such as fillers or binders can be added to adapt the processability of the dried nanoparticle dispersion or to tailor the final dosage form.

Alternatively, the aqueous dispersion of the colloidal particles can be incorporated into the solid dosage form as a liquid, for example by granulation of suitable fillers with the colloidal dispersion to form a granulation. Such granules can subsequently be filled into capsules or be compressed into tablets. Alternatively, through layering of the dispersion onto e.g. sugar-pellets as carriers in a fluidized bed a solid form for nanoparticles can be obtained [Schmidt and Bodmeier, 1999]. These ways of manufacturing tablet cores, or granules or pellets can potentially by followed by a coating step to reveal a film-coated tablet or film-coated granules in a capsule as the final dosage form.

The mechanical stress applied during drying and/or compression of nanoparticle-based formulations has to be considered when selecting a process for transforming a nanoparticle dispersion into a solid dosage form. Processes and process parameters need to take the susceptibility of the nanoparticles in question to stress into consideration. The aim is to maintain the characteristics of the colloidal carriers after formulation into the final dosage form and redispersion therefrom after administration. Potentially harmful are phenomena like fusion or aggregation of the nanoparticles to larger agglomerates, their binding to tabletting excipients, or collapsing of nanocapsules leading to pre-mature release of the content during preparation or storage. All of these occurrences, if happening, do not permit redispersion of a colloidal system after ingestion of the respective solid dosage form. Two critical parameters for the complete redispersibility of nanoparticles were identified as high minimum film formation temperature (MFT) of the polymer dispersion and a good wettability of the dried polymeric nanoparticles [Schmidt and Bodmeier, 1999]. A low MFT led to fusion of the nanoparticles, revealing lumps or a coherent film not dispersing upon contact with fluids; a low wettability hindered the re-generation of the large overall surface area of a dispersion by limited affinity of water to the surface of the dried nanoparticles. Instead, hydrophobic bonds between the nanoparticles held them together as large agglomerates.

Another general aspect requiring thorough investigation is the chemical stability upon storage, which will not be further described here.

A solid dosage form circumventing any compression-related issues was developed by Bodmeier *et al.* [1989]. The nanoparticles were entrapped in beads formed by ionotropic gelation of chitosan with tripolyphosphate or sodium alginate with calcium chloride, respectively. The resulting beads can be filled into capsules for oral administration

Long acting matrix tablets were prepared by direct compression of drug containing PLGA nanoparticles [Murakami *et al.*, 2000]. The tablet showed a biphasic release pattern, which was altered by variation in tablet weight and size, but the amount released per unit surface area remained constant. The release pattern of such a preparation would be based only on the swelling properties of the nanoparticles and shall be independent of the drug contained within.

References

Abdelwahed, W., Degobert, G. and Fessi, H. (2006). Investigation of Nanocapsules Stabilization by Amorphous Excipients during Freeze-drying and Storage. *Eur. J. Pharm. Biopharm.*, 63, pp. 87-94.

Al Khoury-Fallouh, N., Roblot-Treupel, L., Fessi, H., Devissaguet, J.P. and Puisieux, F. (1986). Development of a New Process for the Manufacture of Poly(Isobutylcyanoacrylate) Nanocapsules. *Int. J. Pharm.*, 28, pp. 125-136.

Allémann, E., Gurny, R. and Doelker, E. (1992). Preparation of Aqueous Polymeric Nanodispersions by a Reversible Salting-out Process: Influence of Process Parameters on Particle Size. *Int. J. Pharm.*, 87, pp. 247-253.

Allémann, E., Gurny, R. and Doelker, E. (1993 a). Drug-loaded Nanoparticles - Preparation Methods and Drug Targeting Issues. *Eur. J. Pharm. Biopharm.*, 39, pp. 173-191.

Allémann, E., Leroux J. C., Gurny R. and Doelker E. (1993 b). In vitro extended-release properties of drug-loaded poly(DL-lactic acid) nanoparticles produced by a salting-out procedure. *Pharm. Res.* 10, pp. 1732-1737.

Arica Yegin, B., Benoit, J. P. and Lamprecht, A. (2006). Paclitaxel-loaded Lipid Nanoparticles Prepared by Solvent Injection or Ultrasound Emulsification. *Drug Dev. Ind. Pharm.*, 32, pp. 1089-1094.

Artursson, P., Lindmark, T., Davis, S. S. and Illum, L. (1994). Effect of Chitosan on the Permeability of Monolayers of Intestinal Epithelial Cells (Caco-2). *Pharm. Res.*, 11, pp. 1358-1361.

Attwood, D. (1994). Microemulsions, in: Kreuter, J. (ed.), Colloidal Drug Delivery Systems (Marcel Dekker Inc., New York, Basel, Hong Kong), pp. 31-71.

Auvillain, M., Cavé, G., Fessi, H. and Devissaguet, J.P. (1989). Lyophilisation de Vecteurs Colloidaux Submicroscopiques. *S.T.P. Pharma*, 5, pp. 738-744.

Bangham, A. D., Standish, M. M. and Watkins, J. C. (1965). Diffusion of Univalent Ions across the Lamellae of Swollen Phospholipids. *J. Mol. Biol.*, 13, pp. 238-252.

Bangham, A. D. (1968). Membrane Models with Phospholipids. *Prog. Biophys. Mol. Biol.*, 18, pp. 29-95.

Barenholz, Y., Amselem, S. and Lichtenberg, D. (1979). A New Method for Preparation of Phospholipid Vesicles (liposomes) – french press. *FEBS Lett.*, 99, pp. 210-214.

Barichello, J. M., Morishita, M., Takayama, K. and Nagai, T. (1999). Encapsulation of Hydrophilic and Lipophilic Drugs in PLGA Nanoparticles by the Nanoprecipitation Method. *Drug Dev. Ind. Pharm.*, 25, pp. 471-476.

Batzri, S. and Korn E. D. (1973). Single Bilayer Liposomes Prepared without Sonication. *Biochim. Biophys. Acta*, 298, pp. 1015-1019.

Bazile, D. V., Ropert, C., Huve, P., Verrecchia, T., Marlard, M., Frydman, A., Veillard, M. and Spenlehauer, G. (1992). Body Distribution of Fully Biodegradable [C-14]

Poly(lactic acid) Nanoparticles Coated with Albumin after Parenteral Administration to Rats. *Biomaterials*, 13, pp. 1093-1102.

Bodmeier, R. and Paeratakul, O. (1989). A Novel Approach to the Oral Delivery of Micro- or Nanoparticles. *Pharm. Res.*, 5, pp. 413-417.

Bodmeier, R. and Chen, H. (1990). Indomethacin Polymeric Nanosuspensions Prepared by Microfluidization. *J. Control. Rel.*, 12, pp. 223-233.

Bodmeier, R., Chen, H., Tyle, P. and Jarosz, P. (1991). Spontaneous Formation of Drug-containing Acrylic Nanoparticles. *J. Microencaps.*, 8, pp. 161-170.

Brannon-Peppas, L. (1995). Recent Advances on the Use of Biodegradable Microparticles and Nanoparticles in Controlled Drug Delivery. *Int. J. Pharm.*, 116, pp. 1-9.

Breton, P., Guillon, X., Roy, D., Lescure, F., Riess, G., Bru, N., Roques-Carmes and C. (1998). Physico-chemical Characterization, Preparation and Performance of Poly (methylidene malonate 2.1.2) Nanoparticles. *Biomaterials*, 19, pp. 271-281.

Chattopadhyay, P. and Gupta, R. B. (2001). Production of Griseofulvin Nanoparticles Using Supercritical CO(2) Antisolvent with Enhanced Mass Transfer. *Int. J. Pharm.*, 228, pp. 19-31.

Chen, H. and Langer, R. (1998). Oral Particulate Delivery: Status and Future Trends. *Adv. Drug Del. Rev.*, 34, pp. 339-350.

Chen, X., Young, T. J., Sarkari, M., Williams, R. O., Johnston, K. P. (2002). Preparation of Cyclosporine A Nanoparticles by Evaporation Precipitation into Aqueous Solution. *Int. J. Pharm.*, 242, pp. 3-14.

Chen, J., Zhou, M., Shao, L., Wang, Y., Yun, J., Chew, N. Y. K. and Chan, H. (2004). Feasibility of Preparing Nanodrugs by High-Gravity Reactive Precipitation. *Int. J. Pharm.*, 269, pp. 267-274.

Constantinides, P. P., Scalart, J. P., Lancaster, C., Marcello, J., Marks, G., Ellens, H. and Smith, P. L. (1994). Formulation and Intestinal Absorption Enhancement Evaluation of Water-in-Oil Microemulsions Incorporating Medium-Chain Glycerides. *Pharm. Res.*, 11, pp. 1385-1390.

Constantinides, P. P. (1995). Lipid Microemulsions for Improving Drug Dissolution and Oral Absorption: Physical and Biopharmaceutical Aspects. *Pharm. Res.*, 12, pp. 1561-1572.

Couvreur, P., Kante, B., Roland, M., Guiot, P., Bauduin, P. and Speiser, P. (1979). Polycyanoacrylate Nanocapsules as Potential Lysosomotropic Carriers: Preparation, Morphological and Sorptive Properties. *J. Pharm. Pharmacol.*, 31, pp. 331–332.

Couvreur, P. and Vauthier, C. (1991). Polyalkylcyanoacrylate Nanoparticles as Drug Carrier: Present State and Perspectives. *J. Control. Rel.*, 17, pp. 187-198.

Couvreur, P., Dubernet, C. and Puisieux, F. (1995). Controlled Drug Delivery with Nanoparticles: Current Possibilities and Future Trends. *Eur. J. Pharm. Biopharm.*, 41, pp. 2-13.

Crommelin, D. J. A. and Schreier, H. (1994). Liposomes, in: Kreuter, J. (ed.), Colloidal Drug Delivery Systems (Marcel Dekker Inc., New York, Basel, Hong Kong), pp. 73-190.

Damgé, C., Aprahamian, M., Balboni, G., Hoeltzel, A., Andrieu, V. and Devissaguet, J. P. (1987). Polyalkylcyanoacrylate Nanocapsules Increase the Intestinal Absorption of a Lipophilic Drug. *Int. J. Pharm.*, 36, pp. 121-125.

De Chasteigner, S., Cave, G., Fessi, H., Devissaguet, J.P. and Puisieux, F. (1996). Freeze-Drying of Itraconazole-Loaded Nanosphere Suspension: A Feasibility Study. *Drug Dev. Res.*, 38, pp. 116-124.

De Jaeghere, F., Allémann, E., Feijen, J., Doelker , E. and Gurny, R. (1999). Freeze-Drying of PLA-PEO Nanoparticles – from Basic Principles to Rational Optimization. *Proceed. Int. Symp. Control. Rel. Bioact. Mater.*, 26, pp. 709-710.

De Kruijff, B., Cullis, P. R., Radda, G. K. (1975). Outside-inside Distributions and Sizes of Mixed Phosphatidyl-Cholesterol Vesicles. *Biochim. Biophys. Acta*, 436, pp. 729-740.

Desai, N. P., Soon-Shiong, P., Tao, C., Yang, A., Louie, L., Yoa, Z. and Magdassi, S. (1999). Protein Stabilized Pharmacologically Active Agents, Methods for the Preparation Thereof and Methods for the Use Thereof. US Patent 5,916,596.

Drewe, J., Meier, R., Vonderscher, J., Kiss, D., Posanski, U., Kissel, T. and Gyr, K. (1992). Enhancement of the Oral Absorption of Cyclosporin in Man. *Br. J. Clin. Pharmacol.*, 34, pp. 60-64.

Edlund, U. and Albertsson, A. C. (2003). Polyesters Based on Diacid Monomers. *Adv. Drug Deliv. Rev.*, 55, pp. 585-609.

Eerikainen, H., Watanabe, W., Kauppinen E.I. and Ahonen, P. P. (2003). Aerosol Flow Reactor Method for Synthesis of Drug Nanoparticles. *Eur. J. Pharm. Biopharm.*, 55, pp. 357-360.

Fessi, H., Puisieux, F., Devissaguet, J. P., Ammoury, N. and Benita, S. (1989). Nanocapsule Formation by Interfacial Polymer Deposition Following Solvent Displacement. *Int. J. Pharm.*, 55, pp. R1-R4.

Freitas, C. and Müller, R. H. (1998). Effect of Light and Temperature on Zeta Potential and Physical Stability in Solid Lipid Nanoparticle (SLN™) Dispersions. *Int. J. Pharm.*, 168, pp. 221-229.

Gasco, M. and Trotta, M. (1986). Nanoparticles from Microemulsions. *Int. J. Pharm.*, 29, pp. 267-268.

Govender, T., Stolnik, S., Garnett, M.C., Illum, L. and Davis, S.S. (1999). PLGA Nanoparticles Prepared By Nanoprecipitation: Drug Loading and Release Studies of a Water Soluble Drug. *J. Control. Rel.*, 57, pp. 171-185.

Grangier, J. L., Puygrenier, M., Gautier J. C. and Couvreur P. (1991). Nanoparticles as carriers for growth hormone releasing factor. *J. Control. Rel.*, 15, pp. 3-13.

Gregoriadis, G. (1974). Drug-carrier Potential of Liposomes in Cancer Chemotherapy. *Lancet*, 1, pp. 1313-1316.

Gupta, R. B. (2006 a). Fundamentals of Drug Nanoparticles, in Gupta, R.B., Kompella, U.B. (eds.), Nanoparticle Technology for Drug Delivery (Taylor & Francis, New York, London), pp. 1-19.

Gupta, R. B. (2006 b). Supercritical Fluid Technology for Particle Engineering, in Gupta, R.B., Kompella, U.B. (eds.), Nanoparticle Technology for Drug Delivery, (Taylor & Francis, New York, London), pp. 53-84.

Gurny, R., Peppas, N. A., Harrington, D. D. and Banker, G.S. (1981). Development of Biodegradable and Injectable Latices for Controlled Release of Potent Drugs. *Drug Dev. Ind. Pharm.*, 7, pp. 1-25.

Heurtault, B., Saulnier, P., Pech, B., Proust, J. E., Benoit J.P. (2002). A Novel Phase Inversion-Based Process for the Preparation of Lipid Nanocarriers. *Pharm. Res.*, 19, pp. 875-880.

Heurtault, B., Saulnier, B., Pech, B., Proust, J. P. and Benoit, J. P. (2003). Physico-chemical Stability of Colloidal Lipid Particles. *Biomaterial*, 24, pp. 4283-4300.

Hochman, J. and Artursson, P. (1994). Mechanismus of Absorption Enhancement and Tight Junction Regulation. *J. Control. Rel.*, 29, pp. 253-267.

Hu, J., Johnston, K. P. and Williams, R. O. (2004). Rapid Dissolving High Potency Danazol Powders Produced By Spray Freezing Into Liquid Process. *Int. J. Pharm.*, 271, pp. 145-154.

Huang, C.H. (1969). Studies on Phosphatidylcholine Vesicles. Formation and Physical Characteristics. *Biochemistry*, 8, pp. 344-352.

Ibrahim, H., Bindschaedler, C., Doelker, E., Buri, P. and Gurny, R. (1992). Aqueous Nanodispersions Prepared by a Salting-out Process. *Int. J. Pharm.*, 87, pp. 239-246.

Jaiswal, J., Kumar Gupta, S. and Kreuter, J. (2004). Preparation of Biodegradable Cyclosporine Nanoparticles by High-Pressure Emulsification-Solvent Evaporation Process. *J. Control. Rel.*, 96, pp. 169-178.

Jones, M. C. and Leroux, J. C. (1999). Polymeric micelles - a new generation of colloidal drug carriers. *Eur. J. Pharm. Biopharm.*, 48, pp. 101-111.

Kim, Y. I., Fluckinger, L., Hoffman, M., Lartaud-Idjouadiene, I., Atkinson, J. and Maincent, P. (1997). The Antihypertensive Effect of Orally Administered Nifedipine-loaded Nanoparticles in Spontaneously Hypertensive Rats. *Br. J. Pharmacol.*, 120, pp. 399-404.

Kim, A. I., Akers, M. J. and Nail, S. L. (1998). The Physical State of Mannitol after Freeze-Drying: Effects of Mannitol Concentration, Freezing Rate, on a Noncrystallizing Solute. *J. Pharm. Sci.*, 87, pp. 931-935.

Krause, H.J., Schwartz, A. and Rohdewald, P. (1985). Polylactic Acid Nanoparticles, a Colloidal Drug Delivery System for Lipophilic Drugs. *Int. J. Pharm.*, 27, pp. 145-155.

Kreuter, J. (1983). Phyicochemical characterization of polyacrylic nanoparticles. *Int. J. Pharm.*, 14, pp. 43-49.

Kreuter J. (1983). Evaluation of Nanoparticles as a Drug-Delivery System I: Preparation Methods. *Pharm. Acta Helv.*, 58, pp. 196-209.

Kreuter, J. (1991). Peroral Administration of Nanoparticles. *Adv. Drug Del. Rev.*, 7, pp. 71-86.

Kreuter, J. (1994). Nanoparticles, in: Kreuter, J. (ed.), Colloidal Drug Delivery Systems (Marcel Dekker Inc., New York, Basel, Hongkong), pp. 219-342.

Kumar, M. N. V. R., Bakowsky, U. and Lehr, C. M. (2004). Preparation and characterization of cationic PLGA nanospheres as DNA carriers. *Biomaterials*, 25, pp. 1771-1777.

Lamprecht, A., Bouligand, Y. and Benoit, J.P. (2002). New Lipid Nanocapsules Exhibit Sustained Release Properties for Amiodarone. *J. Control. Rel.*, 84, pp. 59-68.

Lee, V. H., Yamamoto, A. and Kompella, B. U. (1991). Mucosal Penetration Enhancers for Facilitation of Peptide and Protein Drug Absorption. *Crit. Rev. Ther. Drug Carrier Syst.*, 8, pp. 91-192.

Lehmann, K. O. R. (1997). Chemistry and Application Properties of Polymethacrylate Coating Systems, in: McGinity, J.W. (ed.), Aqueous Polymeric Coatings for Pharmaceutical Applications (Marcel Dekker Inc., New York, Basel, Hong Kong), pp. 101-176.

Li, Z., Wen, L., Shao, L. and Chen, J. (2004). Fabrication of Porous Hollow Silica Nanoparticles and Their Applications in Drug Release Control. *J. Contr. Rel.*, 98, pp. 245-254.

Liversidge, G. G., Cundy, K. C., Bishop, J. F. and Czekai, D. A. (1992). Surface modified Drug Nanoparticles. US Patent 5, 145,684.

Maincent, P., LeVerge, R., Sado, P., Couvreur, P. and Devissaguet, J.P. (1986). Disposition Kinetics and Oral Bioavailability of Vincamine-Loaded Polyalkyl Cyanoacrylate Nanoparticles. *J. Pharm. Sci.*, 75, pp. 955-958.

Mayer, L. D., Hope, M. J. and Cullis, P. R. (1986). Vesicles of Variable Sizes Produced by a Rapid Extrusion Procedure. *Biochim. Biophys. Acta*, 858, pp. 161-168.

Merisko-Liversidge, E., Liversidge, G.G. and Cooper, E.R. (2003), Nanosizing: a formulation approach for poorly-water-soluble compounds, *Eur. J. Pharm. Sci.*, 18, pp. 113-120.

Mitra, S., Gaur, U., Ghosh, P. C. and Maitra A. N. (2001). Tumour Targeted Delivery of Encapsulated Dextran-Doxorubicin Conjugate Using Chitosan Nanoparticles as Carrier. *J. Control. Rel.*, 74, pp. 317-323.

Molpeceres, J., Guzman, M., Aberturas, M. R., Chacon, M. and Berges, L. (1996). Application of Central Composite Designs to the Preparation of Polycaprolactone Nanoparticles by Solvent Displacement. *J. Pharm. Sci.*, 85, pp. 206-213.

Molpeceres, J., Aberturas, M. R., Chacón, M., Berges, L. and Guzman, M. (1997). Stability of Cyclosporine-Loaded Poly-ε-Caprolactone Nanoparticles. *J. Microencaps.*, 14, pp. 777-787.

Müller, B. W. (1998 a). Mikroemulsionen als neue Wirkstoff-Trägersysteme, in Müller, R.H., Hildebrandt, G.E., Pharmazeutische Technologie: Moderne Arzneiformen (Wissenschaftl. Verlagsges. Stuttgart), pp. 161-168.

Müller, R. H. and Lucks, J. S. (1996), Arzneistoffträger aus Festen Lipidteilchen (Feste Lipidnanosphären (SLN)), European Patent EP 0,605,497.

Müller, R. H. (1998 b). Feste Lipidnanopartikel (SLN), in: Müller, R.H., Hildebrand, G. (eds.): Pharmazeutische Technologie: Moderne Arzneiformen (Wissenschaftliche Verlagsgesellschaft mbH, Stuttgart), pp. 357-366.

Müller, R. H., Möschwitzer, J. and Bushrab, F. N. (2006). Manufacturing of Nanoparticles by Milling and Homegenization Techniques, in: Gupta, R.B., Kompella, U.B. (eds.), Nanoparticle Technology for Drug Delivery (Taylor & Francis, New York, London), pp. 53-84.

Murakami, H., Kobayashi, M., Takeuchi, H. and Kawashima Y. (2000). Utilization of Poly(DL-lactide-co-glycolide) Nanoparticles for Preparation of Mini-depot Tablets by Direct Compression. *J. Control. Rel.*, 67, pp. 29-36.

Niwa, T., Takeuchi, H., Hino, T., Kunou, N. and Kawashima, Y. (1993). Preparations of Biodegradable Nanospheres of Water-Soluble and Insoluble Drugs with D,L-Lactide/Glycolide Copolymer by a Novel Spontaneous Emulsification Solvent Diffusion Method, and the Drug Release Behavior. *J. Control. Rel.*, 25, pp. 89-98.

Olson, F., Hunt, C., Szoka, F. C., Vail, W. and Papahadjopoulos, D. (1979). Preparation of Liposomes of Defined Size Distribution by Extrusion through Polycarbonate Membranes. *Biochim. Biophys. Acta*, 557, pp. 9-23.

Panyam, J., Dali, M. M., Sahoo, S. K., Ma, W., Chakravarthi, S. S., Amidon, G.L., Levy, R.J. and Labhasetwar, V. (2003). Polymer Degradation and In Vitro Release of a Model Protein from Poly(D,L-lactide-co-glycolide) Nano- and Microparticles. *J. Control. Rel.*, 92, pp. 173-187.

Papahadjopoulos, D. and Vail, W. J. (1978). Incorporation of Macromolecules within Large Unilamellar Vesiclles (LUV). *Ann. N. J. Acad. Sci.*, 308, pp. 259-267.

Pouton, C. W. (1997). Formulation of Self-Emulsifying Drug Delivery Systems. *Adv. Drug Del. Rev.*, 25, pp. 47-58.

Qiu, L. Y. and Bae, Y. H. (2006). Polymer Architecture and Drug Delivery. *Pharm. Res.*, 23, pp. 1-30.

Quintanar-Guerrero, D., Allémann, E., Fessi, H. and Doelker, E. (1998). Preparation Techniques and Mechanisms of Formation of Biodegradable Nanoparticles from Preformed Polymers. *Drug Dev. Ind. Pharm.*, 24, pp. 1113-1128.

Rajaonarivony, M., Vauthier, C., Couarraze, G., Puisieux, F. and Couvreur, P. (1993). Development of a New Drug Carrier Made from Alginate. *J. Pharm. Sci.*, 82, pp. 912-917.

Reverchon, E., De Marco, I. and Della Porta, G. (2002). Rifampicin Microparticles Production by Supercritical Antisolvent Precipitation. *Int. J. Pharm.*, 243, pp. 83-91.

Rogers J. A. and Anderson K. E. (1998). The Potential of Liposomes in Oral Drug Delivery. *Crit. Rev. Ther. Drug Carrier Syst.*, 15, pp. 421-480.

Rosca, I. D., Watari F. and Uo, M. (2004). Microparticle Formation and Its Mechanism in Single and Double Emulsion Solvent Evaporation. *J. Control. Rel.*, 99, pp. 271-280.

Runge, S. A. (1998). Feste Lipidnanopartikel (SLN®) als Kolloidaler Arzneistoffträger für die Orale Applikation von Cyclosporin A, Ph.D. Thesis, FU Berlin.

Rupprecht, H. (1993). Physikalisch-Chemische Grundlagen der Gefriertrockung, in: Essig, D., Oschmann, R. (eds.), Lyophilisation, (Wissenschaftliche Verlagsgesellschaft mbH, Stuttgart), pp. 13-38.

Saez, A., Guzman M., Molpeceres, J. and Aberturas, M.R. (2000). Freeze-drying of Polycaprolactone and Poly(D,L-lactic-glycolic) Nanoparticles Induce Minor Particle Size Changes Affecting the Oral Pharmacokinetics of Loaded Drugs. *Eur. J. Pharm. Biopharm.*, 50, pp. 379-387.

Sakuma, S., Sudo, R., Suzuki, N., Kikuchi, H., Akashi, M. and Hayashi, M. (1999). Mucoadhesion of Polystyrene Nanoparticles Having Surface Hydrophilic Polymeric Chains in the Gastrointestinal Tract. *Int. J. Pharm.*, 177, pp. 161-172.

Schmidt, C. and Bodmeier, R. (1999). Incorporation of polymeric nanoparticles into solid dosage forms. *J. Control. Rel.*, 57, pp. 115–125.

Schroeder, U., Sommerfeld, P, Ulrich, S. and Sabel, B. A. (1998). Nanoparticle Technology For Delivery Of Drugs Across the Blood-Brain Barrier. *J. Pharm. Sci.*, 87, pp. 1305-1307.

Schubert, R. (1998). Liposomen in Arzneimitteln, in: Müller, R.H., Hildebrand, G. (eds.), Pharmazeutische Technologie: Moderne Arzneiformen (Wissenschaftliche Verlagsgesellschaft mbH, Stuttgart), pp. 219-242.

Schubert, M. A. and Müller-Goymann, C. C. (2003). Solvent Injection as a New Approach for Manufacturing Lipid Nanoparticles - Evaluation of the Method and Process Parameters. *Eur. J. Pharm. Biopharm.*, 55, pp. 125-131.

Singh, M. and O'Hagan, D. (1998). The Preparation and Characterization of Polymeric Antigen Delivery Systems for Oral Administration. *Adv. Drug Del. Rev.*, 34, pp. 285-304.

Song, C. X., Labhasetwar, V., Murphy, H., Qu, X., Humphrey, W. R., Shebuski, R. J. and Levy, R. J. (1997). Formulation and Characterization of Biodegradable Nanoparticles for Intravascular Local Drug Delivery. *J. Control. Rel.*, 43, pp. 197-212.

Speiser, P. (1990). Lipidnanopellets als Trägersystem zur peroralen Anwendung. European Patent EP 0,167,825.

Speiser, P. (1998). Nanopartikel, in Müller, R.H., Hildebrandt, G.E., Pharmazeutische Technologie: Moderne Arzneiformen (Wissenschaftl. Verlagsges. Stuttgart), pp. 339-356.

Stainmesse, S., Orecchioni, A.M., Nakache, E., Puisieux, F. and Fessi, H. (1995). Formation and Stabilization of a Biodegradable Polymeric Colloidal Suspension of Nanoparticles. *Colloid Polym. Sci.*, 273, pp. 505-511.

Thompson, T. E., Huang, C. and Litman, B. J. (1974). Bilayers and Biomembranes: Compositional Asymmetries Induced by Surface Curvature, in Moscona, A.A. (ed.), The Cell Surface in Development (Whiley and Sons, New York), pp. 1-16.

van der Lubben, I. M., Verhoef, J. C., Borchard, G. and Junginger, H. E. (2001). Chitosan for Mucosal Vaccination. *Adv. Drug Del. Rev.*, 52, pp. 139-144.

Vandervoort J. and Ludwig, A. (2002). Biocompatible Stabilizers in the Preparation of PLGA nanoparticles: A Factorial Design Study. *Int. J. Pharm.*, 238, pp. 77-92

Wang, W. (1996). Oral Protein Drug Delivery. *J. Drug Target.*, 4, pp. 195-232.

Chapter 2

TRANSPORT ACROSS BIOLOGICAL BARRIERS

Noha Nafee
Department of Pharmaceutics, Alexandria University, Alexandria, Egypt
and
Biopharmaceutics and Pharmaceutical Technology, Saarland University, Saarbrücken, Germany

Vivekanand Bhardwaj
Department of Pharmaceutics, National Institute of Pharmaceutical Education and Research (NIPER), S.A.S. Nagar, Punjab, India
and
Biopharmaceutics and Pharmaceutical Technology, Saarland University, Saarbrücken, Germany

Marc Schneider
Pharmaceutical Nanotechnology, Technology, Saarland University, Saarbrücken, Germany
and
Biopharmaceutics Technology, Saarland University, Saarbrücken, Germany

1. Introduction

Inherently, as the word already anticipates, biological barriers are designed to effectively protect organisms from any kind of alien material. At the same time this strongly limits the use of active substances. This holds especially for modern macromolecular drugs produced by biotechnological techniques which are especially powerful. Unfortunately, most of these drugs lack necessary properties such as stability and solubility as well as their potential to cross biological barriers is small [Alonso, 2004; Pardridge, 2006]; resulting in a weak pharmacokinetic profile [Hidalgo, 2001]. Carrier systems offer an essential advantage in protecting these active substances against degradation and metabolism as well as in modifying the interaction with biological material. Nucleic acids (plasmid DNA and antisense oligonucleotides) are a prominent example of substances that are not able to be delivered to the target site in the body without a suitable carrier. The need for specifically tailored and adapted carrier systems was already pointed out in the advent of gene therapy [Bianco, 2004; Verma and Somia, 1997].

In comparison to bulk material, nano-sized objects are known to change their environmental interaction compared to bulk material. This is, first of all, attributed to their small size, that changes their properties from bulk to surface controlled [Nel *et al.*, 2006]. These modified interactions include changed behavior with respect to biological systems [Labhasetwar, 2005]. It is because of this ability that nanotechnology, although a well known concept for decades, is now being intensively applied to bio-systems, especially for drug delivery. Additionally, the usage of nanoparticles may lead to an improved bypassing of multidrug resistance [Fahmy *et al.*, 2005]. The potential attributed to nano-scale objects and structures is enormous resulting in an increasing number of publications and research done in this field [Sahoo and Labhasetwar, 2003].

The two above mentioned aspects, namely the need to deliver high potency drugs and the altered biological activation of nanoparticulate systems stimulate the idea to combine them and to design the right nano-carrier. This need to be done for each drug class as well as for different cell types [Fahmy *et al.*, 2005; Labhasetwar, 2005; Verma and Somia, 1997].

Size is considered to be of utmost importance and it was demonstrated repeatedly that small particles are more efficiently taken up into cells [Panyam and Labhasetwar, 2003 a] or accumulated in tissue [Jani *et al.* 1990; Lamprecht *et al.*, 2001] than larger particles. Regarding the size and cellular uptake several thresholds are determined. 200 nm sized particles and less are internalized using clathrin-coated pits [Rejman *et al.*, 2004] whereas larger objects are taken up via caveolae membrane invaginations. Other pathways are still under research and not clarified yet [Felberbaum-Corti *et al.* 2003]. Particles as large as 500 nm can be taken up by non-phagocytic cells using an energy-dependent process [Fahmy *et al.*, 2005]. However, the mechanisms are not yet fully known and understood and are therefore still under investigation.

In general, the uptake is phrased as targeted delivery. This can be further subdivided: passive targeting is based on effects such as enhanced permeability and retention (EPR) [Maeda *et al.*, 2001; Matsumura and Maeda, 1986], tumor environment and direct local delivery whereas active targeting makes use of the coupling of a tissue/cell specific marker which leads to localized accumulation of the nanocarriers [Kim and Nie, 2005]. Considering cellular interaction, passive and active processes might be further specified and subdivided paracellular and transcellular route comprising the passive [Salama *et al.*, 2006] and concentration dependent barrier transport and the endocytotic pathways comprising active transport. These mechanisms are based on different aspects for uptake like clathrin-mediated, ligand-activated, non-coated vesicular internalization and phago- and pinocytosis [Steimer *et al.*, 2005; Watson *et al.*, 2005 a]. Many agents that are delivered to cells will also be delivered to compartments outside of the classical endocytic pathway. An example of this is the delivery of certain toxins to the endoplasmic reticulum (ER) via the Golgi apparatus [Sandvig and van Deurs, 2002]. In epithelial tissue the temporal opening of tight junctions increasing passive paracellular transport is an important aspect for modified transport [Ferrari, 2005].

With respect to these mechanisms several other factors are important for the interaction with the biological barrier and finally their uptake behavior. The surface properties of the nanocarrier is as well of interest and determines the interaction between carrier and barrier [Labhasetwar, 2005]. This holds true for both specific and non-specific interactions. In this context, the molecular mechanisms of cellular internalization of nanoparticulate matter plays a key role to design and optimize future carrier systems. Even though size plays a crucial role, other factors such as surface chemistry and charge influence the important molecular routes as well in a non negligible way [Jung *et al.*, 2001; Vila *et al.*, 2004] and need to be considered for the optimal design of the carrier system.

Therefore, in the text below some nanoparticulate systems will be considered and classified according to their type of material. Their potential and their achievements with respect to application will be highlighted.

2. Polymeric Nanoparticles

Polymer-based nanoparticles, especially those prepared with biodegradable polymers have become an important area of research in the field of drug and gene delivery [Panyam and Labhasetwar, 2003 a]. Many biodegradable polymers were used to prepare nanoparticles including poly (lactic acid), PLA; poly (D,L-lactide-co-glycolide), PLGA, poly (ε-caprolactone), gelatin and chitosan [Soppimath *et al.*, 2001]. Nanoparticles can be used to deliver hydrophilic and hydrophobic drugs, proteins, vaccines as well as biological macromolecules via a number of routes. Many nanoparticles ensured efficient targeted delivery to the lymphatic system, arterial wall, lungs, or liver [Brannon-Peppas and Blanchette, 2004]. In addition, nanoparticles can cross the blood brain barrier following the opening of the tight junctions e.g. by hyperosmotic mannitol providing sustained delivery of therapeutic agents for difficult-to-treat diseases like brain tumors [Costantino *et al.*, 2006; Kroll *et al.*, 1998]. Some *in vitro* studies have shown that serum does not affect the intracellular uptake of PLGA nanoparticles and hence they can be administered into systemic circulation without particle aggregation or blockage of fine blood capillaries [Sahoo *et al.*, 2002].

Therefore, the study of the transport of these nanoparticulate carriers through different biological barriers is very intriguing. As already mentioned, uptake of particulate systems could occur through various processes such as phagocytosis, fluid phase pinocytosis or receptor mediated endocytosis. Panyam *et al.* [2002] investigated the uptake and distribution of PLGA nanoparticles in various cell lines. In vascular smooth muscle cells, the nanoparticle internalization was found to be incorporated through fluid phase pinocytosis and in part through clathrin-coated pits [Panyam and Labhasetwar, 2003 b]. The uptake was concentration and time dependent; efficiency decreased at higher doses, suggesting that the uptake pathway is a saturable process. Following their uptake, nanoparticles were transported to primary endosomes then to sorting endosomes [Panyam and Labhasetwar, 2003 a]. A fraction is then sorted out of the cell through recycling endosomes while the remaining fraction is transported to secondary endosomes, which then fuse with lysosomes. In the acidic pH of the endo-lysosomes, charge reversal of the nanoparticles occur due to transfer of proton/hydronium ions from the bulk solution to the particle surface [Watson *et al.*, 2005 b]. This allows stronger electrostatic interactions leading to localized destabilization of the membrane and escape of the nanoparticles to the cytoplasmic compartment. The zeta potential was found to be very sensitive to changing pH values, indicating that the surface density of protonated amino groups and the degree of protonation are reversibly responsive to pH changes [Gan *et al.*, 2005]. On the other hand, polystyrene nanoparticles are unable to escape the endo-lysosomal compartment because they do not exhibit a charge reversal with pH changes [Panyam *et al.*, 2002]. Thus, nanoparticles could be directed to different cell compartments either by the proper choice of the polymers or surface modification of the nanoparticles with cationic polymers like chitosan [Janes *et al.*, 2001; Ravi Kumar *et al.*, 2004], poly ethyleneimine [Bivas-Benita *et al.*, 2004; Trimaille *et al.*, 2003] and poly (2-dimethyl-amino)ethyl methacrylate [Munier *et al.*, 2005].

Once the extracellular concentration of nanoparticles decreases, exocytosis begins. Proteins (e.g. albumin) are responsible for inducing nanoparticle exocytosis. While the drop in intracellular nanoparticle levels could lead to lower efficiency of the encapsulated therapeutic agent, it has to be realized that nanoparticle concentration outside the cell may not fall so rapidly *in vivo*. Thus there could be a constant presence of nanoparticles next to the cells, which might lead to mass transport equilibrium being reached, resulting in higher intracellular nanoparticle levels [Panyam and Labhasetwar, 2003 a].

One of the most important biological barriers to controlled drug or gene delivery is the process of opsonization. This is the process by which a foreign organism or particle becomes covered with opsonin proteins, thereby making it more visible to phagocytic cells [Owens III and Peppas, 2006]. A widely used method to slow the opsonization of nanoparticles is the use of hydrophilic polymers such as PEG, poloxamers and poloxamines, which can block the electrostatic and hydrophobic interaction of opsonin with the particle surface and hence imparts sterically stabilized stealth nanoparticles [Csaba *et al.*, 2006]. The characteristics of this layer; thickness, charge, grafting density, molecular conformation and functional groups, all impact the way in which it interacts with opsonin.

Nanoparticle uptake is a function of their colloidal characteristics. Particle size significantly affects cellular and tissue uptake, and in some cell lines, only the submicron size particles are taken up efficiently but not microparticles (e.g. Hepa 1-6, HepG2, and KLN 205) [Zauner *et al.*, 2001]. The efficiency of uptake of 100 nm size particles was 15-250 fold greater than larger size (1 and 10 μm) microparticles [Desai *et al.*, 1996]. Nanoparticles were able to penetrate throughout the submucosal layers while the larger size microparticles were predominantly localized in the epithelium lining [Desai *et al.*, 1996]. Nevertheless, chitosan nanoparticles are able to be internalized into intestinal, nasal, and ocular epithelial cells [Huang *et al.*, 2004; Janes *et al.*, 2001]. Chitosan is known to be a penetration enhancer in acidic environment towards monostratified and pluristratified epithelia both endowed with and lacking tight junction [Dodane *et al.*, 1999]. The uptake of chitosan nanoparticles seems to be related to the size and the superficial charge: the higher the superficial positive charge, the stronger is the affinity between the nanoparticles and the negatively charged cell membranes and mucus respectively [Huang *et al.*, 2004]. The contact time of the carrier systems with the membrane might increase uptake probability

[El-Shabouri, 2002; Hariharan *et al.*, 2006]. Nevertheless, opposite results were found as well, where negatively charged and neutral particles showed an increased uptake into the Peyer's patches of mice [Shakweh *et al.*, 2005].

Another important barrier which is differently composed is skin which confines percutaneous drug delivery. Only a limited number of polymer-based nanoparticulate carrier systems have been investigated with their respect to their potential to influence transdermal drug delivery. Biodegradable particles were found to enhance the penetration of lipophilic compounds when compared to non-particulate formulations [Alvarez-Roman *et al.*, 2004; Luengo *et al.*, 2006; Shim *et al.*, 2004]. Luengo *et al.* proposed a change in the local environment that leads to an increased partition coefficient of the drug into the horny layer of skin [Luengo *et al.*, 2006]. This localized change of pH is due to the degradation of the particles and the surface chemistry. However, other mechanisms might be involved as well.

3. Polyelectrolyte Complexes (Polyplexes)

It is well recognized that cationized polymers readily form complexes with negatively charged DNA through electrostatic interactions [Dang and Leong, 2006]. This condenses DNA and creates a net positive electric charge under appropriate conditions, which facilitates cell attachment and subsequent internalization by means of endocytosis. Among the cationic polymers used as non-viral vectors PEI [Kircheis *et al.*, 2001], poly (L-lysine) (PLL) [Park *et al.* 2006], collagen [Cohen-Sacks *et al.*, 2004], chitosan [Janes *et al.*, 2001], and trimethyl chitosan [Kean *et al.*, 2005].

In order to promote the internalization of DNA into the cells, several cell receptor ligands have been used to take advantage of receptor mediated endocytosis. Galactose-bound cationized polymers such as galactosylated chitosan [Gao *et al.*, 2005] enable direct delivery and internalization of DNA into the liver through the asialoglycoprotein receptors to which galactose binds [Hashida *et al.*, 2001]. Similarly, selective delivery and internalization of DNA into tumors can be achieved with folate- and transferrin-bound cationized polymers [Qian *et al.*, 2002; Sudimack and Lee, 2000]. Recently, chitosan oligomers were substituted with a trisaccharide branch that targets cell-surface lectins to improve the gene delivery to lungs [Issa *et al.*, 2006]. The results indicated a 10-fold increase in gene expression levels in human liver hepatocytes (HepG2) as well as in human bronchial epithelial cell line (16HBE14o-). Furthermore, *in vitro* and *in vivo* transfection confirmed lectin-mediated uptake [Issa *et al.*, 2006].

On the other hand, modification of cationic polymers with PEG enable DNA uptake because of an prolonged systemic circulation time. Accumulation within tumors or at the sites of inflammation due to characteristic changes in the vasculature including increased vascular permeability and a relative lack of lymph vessels are resulting, the so-called EPR-effect [Maeda *et al.*, 2000].

Further modification of chitosan to improve the transfection efficiency of chitosan-DNA polyplexes are described in details in [Danielsen *et al.*, [2005] and Mansouri *et al.*, [2004]].

4. Liposomes

Cationic liposomes made of cationic lipids improve the transfection efficiency of DNA through the formation of liposomes/DNA complexes or lipoplexes. During lipoplex-mediated transfection by endocytosis, therapeutic molecules are prone to degradation within endosomes or lysosomes [Wattiaux *et al.*, 2000]. Various lipids e.g. dioleoylphosphatidylethanolamine (DOPE), cholesterol, and N- [1-(2,3-dimyristyloxy) propyl]-N,N-dimethyl-N-(2-hydroxyethyl) ammonium bromide are capable of facilitating the endosomal release of DNA by destabilizing the endosome membrane [El Ouahabi *et al.*, 1997].

The requirement for nuclear transport of plasmid DNA poses a significant barrier to effective gene expression following gene therapy using non-viral vectors. Incorporation of viral machinery capable of mediating nuclear transport of exogenous DNA into non-viral vectors might enhance the migration of exogenous DNA into the nucleus. Therefore, to avoid degradation prior to reaching the cytoplasm, fusion mediated delivery systems have been developed. A fusigenic viral liposome with an envelope from HVJ (Sendai virus, a mouse parainfluenza virus) was developed [Kaneda and Tabata, 2006]. The virus contains HN- and F-fusion proteins, which bind to the acetyl-type sialic acid and lipids respectively, inducing membrane fusion. DNA-loaded liposomes can be fused with UV-inactivated HVJ to form fusigenic viral liposomes. It is expected that these carriers might be protected from degradation within the endo-lysosomes and enhance the efficiency of gene transfer. Results showed that about 85% of the oligodeoxynucleotide remained intact in the nucleus of human fibroblasts following administration using HVJ-liposomes, compared to 30% following delivery using Lipofectin.

The so far vast majority of applications is in the field of epicutaneous applied liposomes. The first description is found to be nearly 30 years old [Mezei and Gulasekharam, 1980] and from this time on, the number of publications and work done is growing continuously. It was demonstrated for several drugs, that the provision in liposomes

Table 1 Drugs used with liposmal formulations for skin accumulation.

Drug	Reference
Triamcinolone acetonide	[Mezei and Gulasekharam 1980]
Bethamethasone diproprionate	[Korting *et al.* 1990]
Tretinoin	[Schafer-Korting *et al.* 1994]
Dyphililine	[Touitou *et al.* 1992]
Caffeine	[Touitou *et al.*, 1994]
Tetracaine	[Foldvari, 1994]
Cyclosporine	[Egbaria *et al.*, 1990]
Interferone gamma	[Short *et al.*, 1996]
Testosterone	[Ainbinder and Touitou, 2005]

lead to an increased amout of those active substances in the epidermins and the dermis. Eventhough there are several topical therapeutics on the market, the mechanism of liposomal interaction is still unclear. Generally it is accepted that the liposomes do not penetrate the stratum corneum intact. A fusion of the liposomes with the skin lipids is considered to be most likely. Furthermore, a penetration via hair follicles is discussed [Jung *et al.*, 2006].

5. Lipoplexes

Endocytosis is believed to be the major mechanism for DNA delivery by lipid particles [Torchilin, 2005] and positively charged lipids are most widely explored category for this application [Koltover *et al.*, 1998].

Lipo- and polyplexes do this in different ways. The former fuse with the endosomal membrane employing fusogenic material like DOPE [Felgner *et al.*, 1994]. Another approach is making the liposomes using pH sensitive material so that they dissolve in the acidic environment of endosome [Torchilin *et al.*, 1993]. Other pH sensitive material that has been used include oleyl alcohol [Heeremans *et al.*, 1995] and monostearoyl derivatives of morpholine [W. Rubas, 1986]. Polyplexes employ help from polymers such as polyethyleneinimine PEI which can strongly protonate under the acidic pH in endosome creating a charge gradient resulting in water ingress and following swelling and disintegration of the endosome. [Boussif *et al.*, 1995]. PEI can also be used with lipidic nano-systems, creating hybrid mechanism of action.

For compounds that are unstable in the lysosomal environment, however alternative pathways have to be employed to bypass the endocytotic pathway. This could be done using specific carrier mediated pathways. One example uses trans-activating transcriptional activator (TAT) protein marked liposomal formulations, by which particles as big as 200 nm were found to be translocated intracellularly [Torchilin *et al.*, 2001]. However, the transport is slower possibly due to the size hindrance. The liposomes were found to be accumulating perinuclearly, where they degraded in a time dependent manner. This approach was used to deliver DNA into the cells, in close proximity to the nucleus [Torchilin, *et al.* 2003]. These events have been recorded employing fluorescence microscopy.

6. Solid Lipid Nanoparticles (SLN)

In the early 90's solid lipid nanoparticles were introduced as drug carriers in the pharmaceutical field. In general SLN are composed of physiological solid lipids manufactured by a high pressure homogenisation process.

There are various ways in which lipidic systems can increase the bioavailability of incorporated drugs. But the main mechanism is by getting degraded by lipases to generate active mono and diacylglycerols which can solubilize a poorly soluble drug. These lipids are then taken up after emulsification by the action of bile salts. Degradation and subsequent solubilization is faster for finer droplets provided by the SLN. However they can also be designed not to degrade in the gastro-intestinal tract and to be taken up by the other paracellular and intracellular pathways [Müller *et al.*, 2006].

Rate of intracellular accumulation and cytotoxic activity of doxorubicin-loaded SLN differed among different cell lines; in particular, cells of epithelial origin were found to be more sensitive regarding the cytotoxic activity [Serpe *et al.*, 2006]. SLN offer a new approach to improve the oral bioavailability of poorly soluble drugs. An increase in drug bioavailability is seen due to the ability of SLN to increase the saturation solubility and the rate of solubilisation to a degree high enough not to be offset by the rate of passive diffusion thus maintaining a high concentration gradient [Luo *et al.*, 2006].

The transport barriers would include the scavenging mechanisms in the body represented by the reticulo-endothelial system (RES). SLN of testosterone ^{125}I-labelled histamine derivatives, when intravenously injected via the tail vein of Wistar Unilever rats remain in the blood in contrast to many other colloidal drug carriers. Otherwise significant higher amounts of radioactivity would have been determined in organs of the mononuclear phagocytic system (MPS) such as liver and spleen [Weyhers *et al.*, 2006]. When uptake of free insulin incorporated in wheat germ agglutinin (WGA) modified liposomes and SLN was studied from duodenum, jejunum and ileum of rats in an in situ study, the nanoparticle type and delivery site were found to be important factors with respect to increasing the bioavailability of insulin following oral administration. The proteolytic degradation as well as the epithelial permeability were clearly implicated in terms of regional anatomical and physiological variation for mucosal absorption [Zhang *et al.*, 2006]. Since the structural heterogeneity of these three segments is understood, it is self-evident that besides simple diffusion, other transport phenomena exist.

7. Dendrimers

Dendrimers are branching polymer structures which offer advantage of a high drug loading capacity. There size can be easily regulated by the number of generations and size of the branching groups and thus they can exhibit a very narrow size distribution. Although they have a tendency to aggregate in the stomach [Singh and Florence, 2005], dendrimers are effectively taken up from the gastrointestinal tract and more than 99% of the administered dose can be cleared up within 24 hours [Sakthivel *et al.*, 1999]. In an early study, it was shown that the intestine shows segmental variation in uptake of radioactive iodine marked poly (amidoamine) (PAMAM) dendrimers [Ruedeekorn *et al.*, 2000]. Charged dendrimers showed a concentration dependent uptake, probably due to opening of tight junctions or compromising cell membrane permeability [El-Sayed *et al.*, 2002; Tajarobi *et al.*, 2001].

It was postulated in one of the studies, that G2-NH2 dendrimers were transported across Caco-2 cell monolayers by a combination of paracellular transport and an energy dependent process [El-Sayed *et al.*, 2003]. Also, it was shown that dendrimers are not affected by P-glycoprotein efflux [El-Sayed *et al.*, 2003] and in fact can decrease efflux of susceptible drugs when conjugated to them [D'Emanuele *et al.*, 2004]. The endothelial transport across blood vessels has also been reviewed by Ghandehari *et al.* [Kitchens *et al.*, 2005].

8. Carbon Nanomaterials

Since the discovery of fullerenes or 'buckyballs' (C_{60}) in 1985 [Kroto *et al.*, 1985] and shortly after that, the one of carbon nanotubes (CNT) [Iijima, 1991], several interesting applications are envisaged and their principles are demonstrated [Baughman *et al.*, 2002; Martin and Kohli, 2003].

Even though the CNT aspect ratio resembles the one of asbestos fibers and one expects a certain toxicity, controversial data were obtained in the investigated *in vitro* systems [Cui *et al.*, 2005; Wörle-Knirsch *et al.*, 2006]. However, *in vivo* some acute inflammatory pulmonary effects in rodents were observed [Warheit *et al.*, 2004]. Although these effects are not a result of the CNT per se [Warheit *et al.*, 2004]. In any case the fullerenes show a strong antioxidant ability [Willis, 2004].

The most obvious approach is to use the CNT as hollow carriers and load them with a suitable drug. Pure carbon nanomaterials exhibit an inert surface allowing hydrophobic interactions with the environment. This renders the CNT insoluble in pure water resulting in agglomeration and formation of huge aggregates [Bianco, 2004]. To overcome these problems, CNT surfaces' are modified with hydrophilic groups and charged functional moieties [Isobe *et al.*, 2006]. Good cellular uptake of surface modified CNT [Kam *et al.*, 2004; Pantarotto *et al.*, 2004] has been observed and therefore promised to be a vital route. Several ideas were realized regarding the biological application of water-soluble fullerenes. They were applied in photodynamic therapy [Tokuyama *et al.*, 1993], to inhibit HIV-1 protease [Friedman *et al.*, 1993], and for nuclear medicine [Cagle *et al.*, 1999]. Herein, they are activated using neutron bombardment to form a radionuclide.

Functionalized CNT (f-CNT [Bianco, 2004]) and peptide-modified CNT were investigated on various cell types such as human and murine fibroblasts, keratinocytes and Hela cells. The f-CNT were mainly found in the cytosol whereas the peptide-modified CNT were accumulated in the nucleus [Pantarotto *et al.*, 2004]. Regarding the uptake mechanism different results were reported. In the aforementioned experiments uptake via endocytosis was excluded due to the fact that temperature reduction and the presence of inhibitors did not influence the uptake significantly. However, Kam *et al.* [2004] found a temperature dependent uptake into different cell types such as HL60 cells, Jurkat, Chinese hamster ovary and 3T3 fibroblasts. They claim, that the negatively charged (due to carboxyl groups) CNT still exhibit hydrophobic graphite regions and induce the endocytotic route by non-specific binding [Kam *et al.*, 2004]. These data were confirmed using fluorescent stains for the lysosomes in combination with fluorescently labeled CNT.

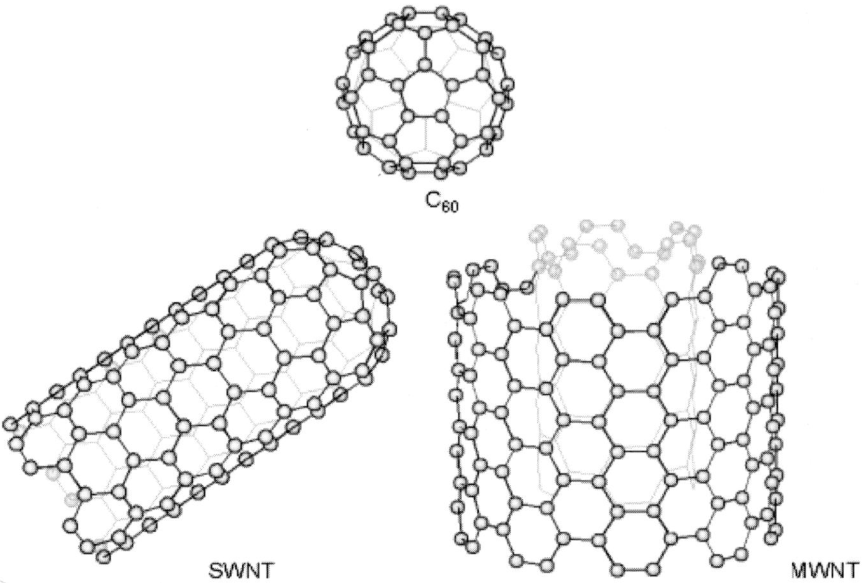

Fig. 1 Image representing a Buckminster fullerene, a single-walled carbon nanotube (SWNT) and a multi-walled carbon nanotube (MWNT). This is a modified version of an image courtesy of Prof. Maruyama, Tokyo, Japan.

A further option is the usage of the hollow cylinder as a transport route [Ito *et al.*, 2003; Cui *et al.*, 2004]. Here objects to be delivered can be transferred through the carbon straw without interacting with the barrier itself. Molecular simulations have already suggested the potential of this approach for water, protons, and oligonucleotides [Gao *et al.*, 2003; Hummer *et al.*, 2001; Kalra *et al.*, 2003].

Another member of the carbon nano-family are carbon nanohorns [Iijima *et al.*, 1999] which have shown to be able to bind and release an anti-inflammatory glucocorticoid (DEX) *in vitro* (Murakami 2004, molecular pharmaceutics) and serve for anticancer drugs [Ajima *et al.*, 2005].

Other tubular structures such as peptide nanotubes [Ghadiri *et al.*, 1993] and self-assembling lipid tubes [Yager and Schoen, 1983] are as well suited for drug delivery purposes and are reviewed in more detail elsewhere [Martin and Kohli, 2003].

9. Metal-based Nanomaterials

Metal-based particulate material is often used for imaging and sensing [Sonvico *et al.*, 2005] as it is well known for gold colloids [Paciotti *et al.*, 2006; Wang *et al.*, 2005; Wang *et al.*, 2002] and others [Cui *et al.* 2004]. In this context peptide-mediated transport across cell membranes was achieved [Tkachenko *et al.*, 2003] and is a promising targeting strategy considering translocation-active peptides such as transportan and penetratin just to name some [Lindgren, *et al.* 2004].

Furthermore magnetic nanoparticles offer exciting possibilities to act as nano-sized drug carriers [Ito *et al.*, 2005] even though their main application is seen in magnetic resonance imaging and cancer thermotherapy [Ito *et al.*, 2005]. Typically, the active substances are attached to the surface either by electrostatic binding, or covalent binding, or other interaction forces [Alexiou *et al.*, 2006; Mehta *et al.*, 1997]. Hereby, the surface is usually pre-coated with hydrophilic polymers to obtain water dispersibility [Dutton *et al.*, 1979; Molday and Mackenzie, 1982; Renshaw *et al.*, 1986; Veiga *et al.*, 2000]. In general, an applied external magnetic field allows to accumulate these particles in the desired body region or tissue [Alexiou *et al.*, 2001; Alexiou *et al.*, 2003]. Nevertheless, the strategy regarding the EPR effect is often utilized for all kinds of nanoparticulate material [Kim and Nie, 2005; Shenoy *et al.*, 2005].

Other non-organic materials are also utilized such as silicon based materials where the payload is attached to the particle's surface [Luo and Saltzman, 2006; Solberg and Landry, 2006; Xu *et al.*, 2006].

An obstacle that need to be considered especially for therapeutic applications is the accumulation of the metal-based particles in the body.

A further applicable system are metal nanoparticles for thermal treatment as shown for gold shells heated up with near infrared (NIR) light [O'Neal *et al.*, 2004]. But these composite materials offer opportunities for drug delivery as well [Hirsch *et al.*, 2006].

10. Quantum Dots (QD)

Quantum dots are semiconductor core fluorescent nanoparticles that have a wide application potential in imaging and diagnosis but not as possible drug carriers so far.

Uptake of QD, like other nano-sized systems can follow passive mechanism of simple endocytosis [Jaiswal *et al.*, 2003], or active uptake by marker tagging to the QD [Chan and Nie, 1998; Chen and Gerion, 2004; Voura *et al.*, 2004].

Some artificial methods that could be used for internalization [Maysinger *et al.*, in press] can be micropipette injection, electroporation, and possibly by induction of plasma membrane damage [Kloepfer *et al.*, 2005].

QD are rather small (2-10 nm) and can be functionalized using antibodies, receptor ligands or receptor targcted peptides to study impart target specificity. Transport of particulate matter of such small size can provide important insights into the details of transport phenomena existing in different types of cells and tissues systems. One very crucial aspect, however, is to ensure that there is no aggregation of the individual dots, without which it would be difficult to arrive at a meaningful result. The exact mechanism of their uptake is still being explored, but they have been imaged in the cytoplasm and even detected inside the nucleus [Jasmina *et al.*, 2005]. QD have been linked to Immunoglobulin G [Wu *et al.*, 2003], Folic acid [Bharali *et al.*, 2005] and Streptavidin [Lidke *et al.*, 2004], and antigens specific for nuclear binding [Hoshino *et al.*, 2004].

11. Other Techniques

Another option, yet still more in the area of basic research is the preparation of hollow shells utilizing the layer-by-layer (LbL)-technology [Decher and Hong, 1991]. This method is characterized by a high flexibility due to the different polymers to be used and has already shown to work as a drug delivery tool when applied to spherical templates [Liu *et al.*, 2005, Skirtach *et al.*, 2006] inter alia using colloidal gold particles for external activation [Skirtach *et al.*, 2006]. Recently, biodegradable micro gels were encapsulated and the incorporated drug were released in pulses [De Geest *et al.*, 2006a, De Geest *et al.*, 2006 b].

As a concluding remark we should stress out, that despite all the promising perspectives of nanotechnology for therapeutic and diagnostic applications, the research need to be accompanied with a proper risk assessment of the nanoparticulate systems impact on the organisms.

References

Ainbinder, D. and Touitou, E. (2005). Testosterone ethosomes for enhanced transdermal delivery. *Drug Deliv.*, 12, pp. 297-303.

Ajima, K., Yudasaka, M., Murakami, T., Maigne, A., Shiba, K. and Iijima, S. (2005). Carbon nanohorns as anticancer drug carriers. *Molecular Pharmaceutics*, 2(6), pp. 475-480.

Alexiou, C., Arnold, W., Hulin, P., Klein, R. J., Renz, H., Parak, F. G., Bergemann, C. and Lübbe, A. S. (2001). Magnetic mitoxantrone nanoparticle detection by histology, X-ray and MRI after magnetic tumor targeting. *J. Magnet. Magnetic Mat.*, 225, pp. 187-193.

Alexiou, C., Jurgons, R., Schmid, R. J., Bergemann, C., Henke, J., Erhardt, W., Huenges, E. and Parak, F. (2003). Magnetic drug targeting - Biodistribution of the magnetic carrier and the chemotherapeutic agent mitoxantrone after locoregional cancer treatment. *J. Drug Target.*, 11, pp. 139-149.

Alexiou, C., Schmid, R. J., Jurgons, R., Kremer, M., Wanner, G., Bergemann, C., Huenges, E., Nawroth, T., Arnold, W. and Parak, F. G. (2006). Targeting cancer cells: Magnetic nanoparticles as drug carriers. *Eur. Biophys. J.*, 35, pp. 446-450.

Alonso, M. J. (2004). Nanomedicines for overcoming biological barriers. *Biomed. Pharmacotherap.*, 58, pp. 168-172.

Alvarez-Roman, R., Naik, A., Kalia, Y. N., Guy, R. H. and Fessi, H. (2004). Enhancement of topical delivery from biodegradable nanoparticles. *Pharm. Res.*, 21, pp. 1818-1825.

Baughman, R. H., Zakhidov, A. A. and De Heer, W. A. (2002). Carbon nanotubes - The route toward applications. *Science*, 297, pp. 787-792.

Bharali, D. J., Lucey, D. W., Jayakumar, H., Pudavar, H. E. and Prasad, P. N. (2005). Folate-Receptor-Mediated Delivery of InP Quantum Dots for Bioimaging Using Confocal and Two-Photon Microscopy. pp. 11364-11371.

Bianco, A. (2004). Carbon nanotubes for the delivery of therapeutic molecules. *Expert Opin. Drug Deliv.*, 1, pp. 57-65.

Bivas-Benita, M., Romeijn, S., Junginger, H. E. and Borchard, G. (2004). PLGA-PEI nanoparticles for gene delivery to pulmonary epithelium. *Eur. J. Pharm. Biopharm.*, 58, pp. 1-6.

Boussif, O., Lezoualc'h, F., Zanta, M. A., Mergny, M. D., Scherman, D., Demeneix, B. and Behr, J. (1995). A Versatile Vector for Gene and Oligonucleotide Transfer into Cells in Culture and in vivo: Polyethylenimine. *Proc. Nat. Acad. Sci.*, 92, pp. 7297-7301.

Brannon-Peppas, L. and Blanchette, J. O. (2004). Nanoparticle and targeted systems for cancer therapy. *Adv. Drug Deliv. Rev.*, 56, pp. 1649-1659.

Cagle, D. W., Kennel, S. J., Mirzadeh, S., Alford, J. M. and Wilson, L. J. (1999). In vivo studies of fullerene-based materials using endohedral metallofullerene radiotracers. *Proc. Nat. Acad. Sci.*, 96, pp. 5182-5187.

Chan, W. C. and Nie, S. (1998). Quantum Dot Bioconjugates for Ultrasensitive Nonisotopic Detection. *Science*, 281, pp. 2016-2018.

Chen, F. and Gerion, D. (2004). Fluorescent CdSe/ZnS Nanocrystal-Peptide Conjugates for Long-term, Nontoxic Imaging and Nuclear Targeting in Living Cells. *Nano Lett.*, 4, pp. 1827-1832.

Cohen-Sacks, H., Elazar, V., Gao, J., Golomb, A., Adwan, H., Korchov, N., Levy, R. J., Berger, M. R. and Golomb, G. (2004). Delivery and expression of pDNA embedded in collagen matrices. *J. Control. Rel.*, 95, pp. 309-320.

Costantino, L., Gandolfi, F., Bossy-Nobs, L., Tosi, G., Gurny, R., Rivasi, F., Angela Vandelli, M. and Forni, F. (2006). Nanoparticulate drug carriers based on hybrid poly(d,l-lactide-co-glycolide)-dendron structures. *Biomaterials*, 27, pp. 4635-4645.

Csaba, N., Sanchez, A. and Alonso, M. J. (2006). PLGA:Poloxamer and PLGA:Poloxamine blend nanostructures as carriers for nasal gene delivery. *J. Control. Rel.*, 113, pp. 164-172.

Cui, D., Ozkan, C. S., Ravindran, S., Kong, Y. and Gao, H. (2004). Encapsulation of Pt-labelled DNA Molecules inside Carbon Nanotubes. *Mech. Chem. Biosystems*, 1, pp. 113-122.

Cui, D., Tian, F., Ozkan, C. S., Wang, M. and Gao, H. (2005). Effect of single wall carbon nanotubes on human HEK293 cells. *Tox. Lett.*, 155, pp. 73-85.

D'Emanuele, A., Jevprasesphant, R., Penny, J. and Attwood, D. (2004). The use of a dendrimer-propranolol prodrug to bypass efflux transporters and enhance oral bioavailability. *Journal of Controlled Release*, 95(3), pp. 447-453.

Dang, J. M. and Leong, K. W. (2006). Natural polymers for gene delivery and tissue engineering. *Adv. Drug Deliv. Rev.*, 58, pp. 487-499.

Danielsen, S., Strand, S., de Lange Davies, C. and Stokke, B. T. (2005). Glycosaminoglycan destabilization of DNA-chitosan polyplexes for gene delivery depends on chitosan chain length and GAG properties. *Biochim. Biophys. Acta*, 1721, pp. 44-54.

De Geest, B. G., Déjugnat, C., Verhoeven, E., Sukhorukov, G. B., Jonas, A. M., Plain, J., Demeester, J. and De Smedt, S. C. (2006 a). Layer-by-Layer coating of degradable microgels for pulsed drug delivery. *J. Control. Rel.*, 116, pp. 159-169.

De Geest, B. G., Stubbe, B. G., Jonas, A. M., Van Thienen, T., Hinrichs, W. L. J., Demeester, J. and De Smedt, S. C. (2006 b). Self-exploding lipid-coated microgels. *Biomacromolecules*, 7, pp. 373-379.

Decher, G. and Hong, J. D. (1991). Buildup of Ultrathin Multilayer Films by a Self-Assembly Process 2. Consecutive Adsorption of Anionic and Cationic Bipolar Amphiphiles and Polyelectrolytes on Charged Surfaces. *Phys. Chem. Chem. Phys.*, 95, pp. 1430-1434.

Desai, M. P., Labhasetwar, V., Amidon, G. L. and Levy, R. J. (1996). Gastrointestinal uptake of biodegradable microparticles, pp. Effect of particle size. *Pharm. Res.*, 13, pp. 1838-1845.

Dodane, V., Amin Khan, M. and Merwin, J. R. (1999). Effect of chitosan on epithelial permeability and structure. *I. J. Pharm.*, 182, pp. 21-32.

Dutton, A. H., Tokuyasu, K. T. and Singer, S. J. (1979). Iron-dextran antibody conjugates, pp. General method for simultaneous staining of two components in high-resolution immunoelectron microscopy. *Proc. Nat. Acad. Sci.*, 76, pp. 3392-3396.

Egbaria, K., Ramachandran, C. and Weiner, N. (1990). Topical delivery of ciclosporin, pp. Evaluation of various formulations using in vitro diffusion studies in hairless mouse skin. *Skin Pharmacol.*, 3, pp. 21-28.

El-Sayed, M., Ginski, M., Rhodes, C. and Ghandehari, H. (2002). Transepithelial transport of poly(amidoamine) dendrimers across Caco-2 cell monolayers. *J. Control. Rel.*, 81, pp. 355-365.

El-Sayed, M., Rhodes, C. A., Ginski, M. and Ghandehari, H. (2003). Transport mechanism(s) of poly (amidoamine) dendrimers across Caco-2 cell monolayers. *Int. J. Pharm.*, 265, pp. 151-157.

El-Shabouri, M. H. (2002). Positively charged nanoparticles for improving the oral bioavailability of cyclosporin-A. *Int. J. Pharm.*, 249, pp. 101-108.

El Ouahabi, A., Thiry, M., Pector, V., Fuks, R., Ruysschaert, J. M. and Vandenbranden, M. (1997). The role of endosome destabilizing activity in the gene transfer process mediated by cationic lipids. *FEBS Letters*, 414, pp. 187-192.

Fahmy, T. M., Fong, P. M., Goyal, A. and Saltzman, W. M. (2005). Targeted for drug delivery. *Materials Today*, 8, pp. 18-26.

Felberbaum-Corti, M., Van Der Goot, F. G. and Gruenberg, J. (2003). Sliding doors: Clathrin-coated pits or caveolae? *Nat. Cell Biol.*, 5, pp. 382-384.

Felgner, J. H., Kumar, R., Sridhar, C. N., Wheeler, C. J., Tsai, Y. J., Border, R., Ramsey, P., Martin, M. and Felgner, P. L. (1994). Enhanced gene delivery and mechanism studies with a novel series of cationic lipid formulations. *J. Biol. Chem.*, 269, pp. 2550-2561.

Ferrari, M. (2005). Cancer nanotechnology: Opportunities and challenges. *Nat. Rev. Cancer*, 5, pp. 161-171.

Foldvari, M. (1994). In vitro cutaneous and percutaneous delivery and in vivo efficacy of tetracaine from liposomal and conventional vehicles. *Pharm. Res.*, 11, pp. 1593-1598.

Friedman, S. H., DeCamp, D. L., Sijbesma, R. P., Srdanov, G., Wudl, F. and Kenyon, G. L. (1993). Inhibition of the HIV-1 protease by fullerene derivatives: Model building studies and experimental verification. *J. Am. Chem. Soc.*, 115, pp. 6506-6509.

Gan, Q., Wang, T., Cochrane, C. and McCarron, P. (2005). Modulation of surface charge, particle size and morphological properties of chitosan-TPP nanoparticles intended for gene delivery. *Coll. Surf. B: Biointerf.*, 44, pp. 65-73.

Gao, H., Kong, Y., Cui, D. and Ozkan, C. S. (2003). Spontaneous insertion of DNA oligonucleotides into carbon nanotubes. *Nano Lett.*, 3, pp. 471-473.

Gao, S., Chen, J., Dong, L., Ding, Z., Yang, Y.-h. and Zhang, J. (2005). Targeting delivery of oligonucleotide and plasmid DNA to hepatocyte via galactosylated chitosan vector. *Eur. J. Pharm. Biopharm.*, 60, pp. 327-334.

Ghadiri, M. R., Granja, J. R., Milligan, R. A., McRee, D. E. and Khazanovich, N. (1993). Self-assembling organic nanotubes based on a cyclic peptide architecture. *Nature*, 366, pp. 324-327.

Hariharan, S., Bhardwaj, V., Bala, I., Sitterberg, J., Bakowsky, U. and Ravi Kumar, M. N. V. (2006). Design of estradiol loaded PLGA nanoparticulate formulations: A potential oral delivery system for hormone therapy. *Pharm. Res.*, 23, pp. 184-195.

Hashida, M., Nishikawa, M., Yamashita, F. and Takakura, Y. (2001). Cell-specific delivery of genes with glycosylated carriers. *Adv. Drug Deliv. Rev.*, 52, pp. 187-196.

Heeremans, J. L. M., Prevost, R., Bekkers, M. E. A., Los, P., Emeis, J. J., Kluft, C. and Crommelin, D. J. A. (1995). Thrombolytic treatment with tissue-type plasminogen activator (t-PA) containing liposomes in rabbits: A comparison with free t-PA. *Thromb. Haemost.*, 73, pp. 488-494.

Hidalgo, I. J. (2001). Assessing the absorption of new pharmaceuticals. *Curr. Topics Med. Chem.* 1, pp. 385-401.

Hirsch, L. R., Gobin, A. M., Lowery, A. R., Tam, F., Drezek, R. A., Halas, N. J. and West, J. L. (2006). Metal nanoshells. *Ann. Biomed. Engin.*, 34, pp. 15-22.

Hoshino, A., Fujioka, K., Oku, T., Nakamura, S., Suga, M., Yamaguchi, Y., Suzuki, K., Yasuhara, M. and Yamamoto, K. (2004). Quantum dots targeted to the assigned organelle in living cells. *Microbiol. Immun.*, 48, pp. 985-994.

Huang, M., Khor, E. and Lim, L.-Y. (2004). Uptake and cytotoxicity of chitosan molecules and nanparticles: effects of molecular weight and degree of deacetylation. *Pharm. Res.*, 21, pp. 344-353.

Hummer, G., Rasaiah, J. C. and Noworyta, J. P. (2001). Water conduction through the hydrophobic channel of a carbon nanotube. *Nature*, 414, pp. 188-190.

Iijima, S. (1991). Helical microtubules of graphitic carbon. *Nature*, 354, pp. 56-58.

Iijima, S., Yudasaka, M., Yamada, R., Bandow, S., Suenaga, K., Kokai, F. and Takahashi, K. (1999). Nano-aggregates of single-walled graphitic carbon nano-horns. *Chem. Phys. Lett.*, 309, pp. 165-170.

Isobe, H., Nakanishi, W., Tomita, N., Jinno, S., Okayama, H. and Nakamura, E. (2006). Gene delivery by aminofullerenes: Structural requirements for efficient transfection. *Chemistry - An Asian Journal*, 1, pp. 167-175.

Issa, M. M., Koping-Hoggard, M., Tommeraas, K., Varum, K. M., Christensen, B. E., Strand, S. P. and Artursson, P. (2006). Targeted gene delivery with trisaccharide-substituted chitosan oligomers in vitro and after lung administration in vivo. *J. Control. Rel.*, 115, pp. 103-112.

Ito, A., Shinkai, M., Honda, H. and Kobayashi, T. (2005). Medical Application of Functionalized Magnetic Nanoparticles. *J. Biosci. Bioengin.*, 100, pp. 1-11.

Ito, T., Sun, L. and Crooks, R. M. (2003). Observation of DNA transport through a single carbon nanotube channel using fluorescence microscopy. *Chem. Comm.*, 9, pp. 1482-1483.

Jaiswal, J. K., Mattoussi, H., Mauro, J. M. and Simon, S. M. (2003). Long-term multiple color imaging of live cells using quantum dot bioconjugates. *Nat. Biotech.*, 21, pp. 47-51.

Janes, K. A., Calvo, P. and Alonso, M. J. (2001). Polysaccharide colloidal particles as delivery systems for macromolecules. *Adv. Drug Deliv. Rev.*, 47, pp. 83-97.

Jani, P., Halbert, G. W., Langridge, J. and Florence, A. T. (1990). Nanoparticle uptake by the rat gastrointestinal mucosa: Quantitation and particle size dependency. *J. Pharm. Pharmacol.*, 42, pp. 821-826.

Jasmina, L., Hassan, S. B., Yan, C., Genevieve, R. A. F., FranÃ§oise, M. W. and Dusica, M. (2005). Differences in subcellular distribution and toxicity of green and red emitting CdTe quantum dots. *J. Mol. Med.*, V83, pp. 377-385.

Jung, S., Otberg, N., Thiede, G., Richter, H., Sterry, W., Panzner, S. and Lademann, J. (2006). Innovative liposomes as a transfollicular drug delivery system: Penetration into porcine hair follicles. *J. Invest. Dermatol.*, 126, pp. 1728-1732.

Jung, T., Kamm, W., Breitenbach, A., Hungerer, K.-D., Hundt, E. and Kissel, T. (2001). Tetanus toxoid loaded nanoparticles from sulfobutylated poly(vinyl alcohol)-graft-poly(lactide-co-glycolide): Evaluation of antibody response after oral and nasal application in mice. *Pharm. Res.*, 18, pp. 352-360.

Kalra, A., Garde, S. and Hummer, G. (2003). Osmotic water transport through carbon nanotube membranes. *Proc. Nat. Acad. Sci.*, 100, pp. 10175-10180.

Kam, N. W. S., Jessop, T. C., Wender, P. A. and Dai, H. J. (2004). Nanotube molecular transporters: Internalization of carbon nanotube-protein conjugates into mammalian cells. *J. Am. Chem. Soc.*, pp. 6850-6851.

Kaneda, Y. and Tabata, Y. (2006). Non-viral vectors for cancer therapy. *Cancer Sci.*, 97, pp. 348-354.

Kean, T., Roth, S. and Thanou, M. (2005). Trimethylated chitosans as non-viral gene delivery vectors: Cytotoxicity and transfection efficiency. *J. Control. Rel.*, 103, pp. 643-653.

Kim, G. J. and Nie, S. (2005). Targeted cancer nanotherapy. *Materials Today*, 8, pp. 28-33.

Kircheis, R., Wightman, L. and Wagner, E. (2001). Design and gene delivery activity of modified polyethylenimines. *Adv. Drug Deliv. Rev.*, 53, pp. 341-358.

Kitchens, K. M., El-Sayed, M. E. H. and Ghandehari, H. (2005). Transepithelial and endothelial transport of poly (amidoamine) dendrimers. *Adv. Drug Deliv. Rev.*, 57, pp. 2163-2176.

Kloepfer, J. A., Mielke, R. E. and Nadeau, J. L. (2005). Uptake of CdSe and CdSe/ZnS Quantum Dots into Bacteria via Purine-Dependent Mechanisms. *Appl. Environ. Microbiol.*, 71, pp. 2548-2557.

Koltover, I., Salditt, T., Radler, J. O. and Safinya, C. R. (1998). An Inverted Hexagonal Phase of Cationic Liposome-DNA Complexes Related to DNA Release and Delivery. *Science*, 281, pp. 78-81.

Korting, H. C., Zienicke, H., Schafer-Korting, M. and Braun-Falco, O. (1990). Liposome encapsulation improves efficacy of betamethasone dipropionate in atopic eczema but not in psoriasis vulgaris. *Eur. J. Clin. Pharmacol.*, 39, pp. 349-351.

Kroll, R. A., Pagel, M. A., Muldoon, L. L., Roman-Goldstein, S., Fiamengo, S. A. and Neuwelt, E. A. (1998). Improving drug delivery to intracerebral tumor and surrounding brain in a rodent model: a comparison of osmotic versus bradykinin modification of the blood brain and/or blood tumor barriers. *Neurosurgery*, 43, pp. 879-886.

Kroto, H. W., Heath, J. R., O'Brien, S. C., Curl, R. F. and Smalley, R. E. (1985). C-60-Buckminsterfullerene. *Nature*, 318, pp. 162-163.

Labhasetwar, V. (2005). Nanotechnology for drug and gene therapy: The importance of understanding molecular mechanisms of delivery. *Curr. Opin. Biotech.*, 16, pp. 674-680.

Lamprecht, A., Schäfer, U. and Lehr, C.-M. (2001). Size-dependent bioadhesion of micro- and nanoparticulate carriers to the inflamed colonic mucosa. *Pharm. Res.*, 18, pp. 788-793.

Lidke, D. S., Nagy, P., Heintzmann, R., Arndt-Jovin, D. J., Post, J. N., Grecco, H. E., Jares-Erijman, E. A. and Jovin, T. M. (2004). Quantum dot ligands provide new insights into erbB/HER receptor-mediated signal transduction. *Nat. Biotech.*, 22, pp. 198-203.

Lindgren, M. E., Hallbrink, M. M., Elmquist, A. M. and Langel, U. (2004). Passage of cell-penetrating peptides across a human epithelial cell layer in vitro. *Biochem. J.*, 377, pp. 69-76.

Liu, X., Gao, C., Shen, J. and Mo?hwald, H. (2005). Multilayer microcapsules as anti-cancer drug delivery vehicle: Deposition, sustained release, and in vitro bioactivity. *Macromol. Biosci.*, 5, pp. 1209-1219.

Luengo, J., Weiss, B., Schneider, M., Ehlers, A., Stracke, F., König, K., Kostka, K.-H., Lehr, C.-M. and Schaefer, U. F. (2006). Influence of Nanoencapsulation on Human Skin Transport of Flufenamic Acid. *Skin Pharmacol. Physiol.*, 19, pp. 190-197.

Luo, D. and Saltzman, W. M. (2006). Nonviral gene delivery: Thinking of silica. *Gene Ther.*, 13, pp. 585-586.

Luo, Y., Chen, D., Ren, L., Zhao, X. and Qin, J. (2006). Solid lipid nanoparticles for enhancing vinpocetine's oral bioavailability. *J. Control. Rel.*, 114, pp. 53-59.

Maeda, H., Sawa, T. and Konno, T. (2001). Mechanism of tumor-targeted delivery of macromolecular drugs, including the EPR effect in solid tumor and clinical overview of the prototype polymeric drug SMANCS. *J. Control. Rel.*, 74, pp. 47-61.

Maeda, H., Wu, J., Sawa, T., Matsumura, Y. and Hori, K. (2000). Tumor vascular permeability and the EPR effect in macromolecular therapeutics: a review. *J. Control. Rel.*, 65, pp. 271-284.

Mansouri, S., Lavigne, P., Corsi, K., Benderdour, M., Beaumont, E. and Fernandes, J. C. (2004). Chitosan-DNA nanoparticles as non-viral vectors in gene therapy: strategies to improve transfection efficacy. *Eur. J. Pharm. Biopharm.*, 57, pp. 1-8.

Martin, C. R. and Kohli, P. (2003). The emerging field of nanotube biotechnology. *Nat. Rev. Drug Discov.*, 2, pp. 29-37.

Matsumura, Y. and Maeda, H. (1986). A new concept for macromolecular therapeutics in cancer chemotherapy: Mechanism of tumoritropic accumulation of proteins and the antitumor agent smancs. *Cancer Res.*, 46, pp. 6387-6392.

Maysinger, D., Lovric, J., Eisenberg, A. and Savic, R. (2007). Fate of micelles and quantum dots in cells. *Eur. J. Pharm. Biopharm.*, 65, pp. 270-281.

Mehta, R. V., Upadhyay, R. V., Charles, S. W. and Ramchand, C. N. (1997). Direct binding of protein to magnetic particles. *Biotech. Techn.*, 11, pp. 493-496.

Mezci, M. and Gulasekharam, V. (1980). Liposomes - a selective drug delivery system for the topical route of administration. I. Lotion dosage form. *Life Sci.*, 26, pp. 1473-1477.

Molday, R. S. and Mackenzie, D. (1982). Immunospecific ferromagnetic iron-dextran reagents for the labeling and magnetic separation of cells. *J. Immun. Methods*, 52(3), pp. 353-367.

Mueller, R. H., Runge, S., Ravelli, V., Mehnert, W., Thunemann, A. F. and Souto, E. B. (2006). Oral bioavailability of cyclosporine: Solid lipid nanoparticles (SLN®) versus drug nanocrystals. *Int. J. Pharm.*, 317, pp. 82-89.

Munier, S., Messai, I., Delair, T., Verrier, B. and Ataman-Onal, Y. (2005). Cationic PLA nanoparticles for DNA delivery: Comparison of three surface polycations for DNA binding, protection and transfection properties. *Colloids and Surfaces B: Biointerfaces*, 43, pp. 163-173.

Nel, A., Xia, T., Madler, L. and Li, N. (2006). Toxic Potential of Materials at the Nanolevel. *Science*, 311, pp. 622-627.

O'Neal, D. P., Hirsch, L. R., Halas, N. J., Payne, J. D. and West, J. L. (2004). Photo-thermal tumor ablation in mice using near infrared-absorbing nanoparticles. *Cancer Lett.*, 209, pp. 171-176.

Owens III, D. E. and Peppas, N. A. (2006). Opsonization, biodistribution, and pharmacokinetics of polymeric nanoparticles. *Int. J. Pharm.*, 307, pp. 93-102.

Paciotti, G. F., Kingston, D. G. I. and Tamarkin, L. (2006). Colloidal gold nanoparticles: A novel nanoparticle platform for developing multifunctional tumor-targeted drug delivery vectors. *Drug Dev. Res.*, 67, pp. 47-54.

Pantarotto, D., Briand, J. P., Prato, M. and Bianco, A. (2004). Translocation of bioactive peptides across cell membranes by carbon nanotubes. *Chem. Comm.*, 1, pp. 16-17.

Panyam, J. and Labhasetwar, V. (2003a). Biodegradable nanoparticles for drug and gene delivery to cells and tissue. *Adv. Drug Deliv. Rev.*, 55, pp. 329-347.

Panyam, J. and Labhasetwar, V. (2003b). Dynamics of endocytosis and exocytosis of poly (D,L-lactide-co-glycolide) nanopartilces in vascular smooth muscle cells. *Pharm. Res.*, 20, pp. 110-118.

Panyam, J., Zhou, W.-Z., Prabha, S., Sahoo, S. K. and Labhasetwar, V. (2002). Rapid endo-lysosomal escape of poly(DL-lactide-co-glycolide) nanoparticles: Implications for drug and gene delivery. *FASEB Journal*, 16, pp. 1217-1226.

Pardridge, W. M. (2006). Molecular Trojan horses for blood-brain barrier drug delivery. *Curr. Opin. Pharmacol.*, 6, pp. 494-500.

Park, T. G., Jeong, J. H. and Kim, S. W. (2006). Current status of polymeric gene delivery systems. *Adv. Drug Deliv. Rev.*, 58, pp. 467-486.

Qian, Z. M., Li, H., Sun, H. and Ho, K. (2002). Targeted drug delivery via the transferrin receptor-mediated endocytosis pathway. *Pharmacol. Rev.*, 54, pp. 561-587.

Ravi Kumar, M. N. V., Mohapatra, S. S., Kong, X., Jena, P. K., Bakowsky, U. and Lehr, C.-M. (2004). Cationic poly(lactide-co-glycolide) nanoparticles as efficient in vivo gene transfection agents. *J. Nanosci. Nanotech.*, 4, pp. 1-5.

Rejman, J., Oberle, V., Zuhorn, I. S. and Hoekstra, D. (2004). Size-dependent internalization of particles via the pathways of clathrin-and caveolae-mediated endocytosis. *Biochem. J.*, 377, pp. 159-169.

Renshaw, P. F., Owen, C. S. and McLaughlin, A. C. (1986). Ferromagnetic contrast agents: A new approach. *Magn. Res. Med.*, 3, pp. 217-225.

Ruedeekorn, W., Begoña C. G., Navid, M. and Ruth, D. (2000). Anionic PAMAM Dendrimers Rapidly Cross Adult Rat Intestine In Vitro: A Potential Oral Delivery System. *Pharm. Res.*, 17, pp. 991-998.

Sahoo, S. K. and Labhasetwar, V. (2003). Nanotech approaches to drug delivery and imaging. *Drug Discov. Today*, 8, pp. 1112-1120.

Sahoo, S. K., Panyam, J., Prabha, S. and Labhasetwar, V. (2002). Residual polyvinyl alcohol associated with poly (,-lactide-co-glycolide) nanoparticles affects their physical properties and cellular uptake. *J. Control. Rel.*, 82, pp. 105-114.

Sakthivel, T., Toth, I. and Florence, A. T. (1999). Distribution of a lipidic 2.5 nm diameter dendrimer carrier after oral administration. *Int. J. Pharm.*, 183, pp. 51-55.

Salama, N. N., Eddington, N. D. and Fasano, A. (2006). Tight junction modulation and its relationship to drug delivery. *Adv. Drug Deliv. Rev.*, 58, pp. 15-28.

Sandvig, K. and van Deurs, B. (2002). Membrane traffic exploited by protein toxins. Annu. Rev. Cell Dev. Biol., 18, pp. 1-24.

Schafer-Korting, M., Korting, H. C. and Ponce-Poschl, E. (1994). Liposomal tretinoin for uncomplicated acne vulgaris. *Clin. Invest.*, 72(12), pp. 1086-1091.

Serpe, L., Guido, M., Canaparo, R., Muntoni, E., Cavalli, R., Panzanelli, P., Pepa, C. D., Bargoni, A., Mauro, A., Gasco, M. R., Eandi, M., and Zara, G. P. (2006). Intracellular accumulation and cytotoxicity of doxorubicin with different pharmaceutical formulations in human cancer cell lines. *J. Nanosci. Nanotech.*, 6, pp. 3062-3069.

Shenoy, D., Little, S., Langer, R. and Amiji, M. (2005). Poly(ethylene oxide)-modified poly(amino ester) nanoparticles as a pH-sensitive system for tumor-targeted delivery of hydrophobic drugs: Part 2. In vivo distribution and tumor localization studies. *Pharm. Res.*, 22, pp. 2107-2114.

Shim, J., Kang, H. S., Park, W.-S., Han, S.-H., Kim, J. and Chang, I.-S. (2004). Transdermal delivery of mixnoxidil with block copolymer nanoparticles. *J. Control. Rel.*, 97, pp. 477-484.

Short, S. M., Paasch, B. D., Turner, J. H., Weiner, N., Daugherty, A. L. and Mrsny, R. J. (1996). Percutaneous absorption of biologically-active interferon-gamma in a human skin graft-nude mouse model. *Pharm. Res.*, 13, pp. 1020-1027.

Singh, B. and Florence, A. T. (2005). Hydrophobic dendrimer-derived nanoparticles. *Int. J. Pharm.*, 298, pp. 348-353.

Skirtach, A. G., Munoz Javier, A., Kreft, O., Köhler, K., Piera Alberola, A., Möhwald, H., Parak, W. J. and Sukhorukov, G. B. (2006). Laser-induced release of encapsulated materials inside living cells. *Angew. Chem.*, 45, pp. 4612-4617.

Solberg, S. M. and Landry, C. C. (2006). Adsorption of DNA into mesoporous silica. *J. Phys. Chem., B* 110, pp. 15261-15268.

Sonvico, F., Dubernet, C., Colombo, P. and Couvreur, P. (2005). Metallic colloid nanotechnology, applications in diagnosis and therapeutics. *Curr. Pharm. Design*, 11, pp. 2091-2105.

Soppimath, K. S., Aminabhavi, T. M., Kulkarni, A. R. and Rudzinski, W. E. (2001). Biodegradable polymeric nanoparticles as drug delivery devices. *J. Control. Rel.*, 70(1-2), pp. 1-20.

Steimer, A., Haltner, E. and Lehr, C.-M. (2005). Cell culture models of the respiratory tract relevant to pulmonary drug delivery. *J. Aerosol Med.*, 18, pp. 137-182.

Sudimack, J. and Lee, R. J. (2000). Targeted drug delivery via the folate receptor. *Adv. Drug Deliv. Rev.*, 41(2), pp. 147-162.

Tajarobi, F., El-Sayed, M., Rege, B. D., Polli, J. E. and Ghandehari, H. (2001). Transport of poly amidoamine dendrimers across Madin-Darby canine kidney cells. *Int. J. Pharm.*, 215(1-2), pp. 263-267.

Tkachenko, A. G., Xie, H., Coleman, D., Glomm, W., Ryan, J., Anderson, M. F., Franzen, S. and Feldheim, D. L. (2003). Multifunctional gold nanoparticle-peptide complexes for nuclear targeting. *J. Am. Chem. Soc.*, 125, pp. 4700-4701.

Tokuyama, H., Yamago, S., Nakamura, E., Shiraki, T. and Sugiura, Y. (1993). Photoinduced biochemical activity of fullerene carboxylic acid. *J. Am. Chem. Soc.*, 115(17), pp. 7918-7919.

Torchilin, V. P. (2005). Fluorescence microscopy to follow the targeting of liposomes and micelles to cells and their intracellular fate. *Adv. Drug Deliv. Rev.*, 57(1), pp. 95-109.

Torchilin, V. P., Levchenko, T. S., Rammohan, R., Volodina, N., Papahadjopoulos-Sternberg, B. and D'Souza, G. G. M. (2003). Cell transfection in vitro and in vivo with nontoxic TAT peptide-liposome-DNA complexes. *Proc. Natl. Acad. Sci.*, 100, pp. 1972-1977.

Torchilin, V. P., Rammohan, R., Weissig, V. and Levchenko, T. S. (2001). TAT peptide on the surface of liposomes affords their efficient intracellular delivery even at low temperature and in the presence of metabolic inhibitors. *Proc. Natl. Acad. Sci.*, 98, pp. 8786-8791.

Torchilin, V. P., Zhou, F. and Huang, L. (1993). pH-Sensitive liposomes. *J. Liposome Res.*, 3, pp. 201-255.

Touitou, E., Levi-Schaffer, F., Dayan, N., Alhaique, F. and Riccieri, F. (1994). Modulation of caffeine skin delivery by carrier design: Liposomes versus permeation enhancers. *Int. J. Pharm.*, 103(2), pp. 131-136.

Touitou, E., Shaco-Ezra, N., Dayan, N., Jushynski, M., Rafaeloff, R. and Azoury, R. (1992). Dyphylline liposomes for delivery to the skin. *J. Pharm. Sci.*, 81, pp. 131-134.

Trimaille, T., Pichot, C. and Delair, T. (2003). Surface functionalization of poly(,-lactic acid) nanoparticles with poly(ethylenimine) and plasmid DNA by the layer-by-layer approach. *Colloids and Surfaces A*, 221, pp. 39-48.

Veiga, V., Ryan, D. H., Sourty, E., Llanes, F. and Marchessault, R. H. (2000). Formation and characterization of superparamagnetic cross-linked high amylose starch. *Carbohydrate Polymers*, 42, pp. 353-357.

Verma, I. M. and Somia, N. (1997). Gene therapy - Promises, problems and prospects. *Nature* 389, pp. 239-242.

Vila, A., Gill, H., McCallion, O. and Alonso, M. J. (2004). Transport of PLA-PEG particles across the nasal mucosa: Effect of particle size and PEG coating density. *J. Control. Rel.*, 98(2), pp. 231-244.

Voura, E. B., Jaiswal, J. K., Mattoussi, H. and Simon, S. M. (2004). Tracking metastatic tumor cell extravasation with quantum dot nanocrystals and fluorescence emission-scanning microscopy. *Nat. Med.*, 10(9), pp. 993-998.

Rubas, W., Supersaxo, A., Weder, H. G., Hartmann, H. R., Hengartner, H., Schott, H. and Schwendener, R. (1986). Treatment of murine L1210 lymphoid leukemia and melanoma B16 with lipophilic cytosine arabinoside prodrugs incorporated into unilamellar liposomes. *Int. J. Concer*, 37 pp. 149-154.

Wang, H., Cheng, J.-X., Huff, T. B., Zweifel, D. A., He, W., Low, P. S. and Wei, A. (2005). In vitro and in vivo two-photon luminescence imaging of single gold nanorods. *Proc. Natl. Acad. Sci.*, 102, pp. 15752-15756.

Wang, Q., Lin, T., Tang, L., Johnson, J. E. and Finn, M. G. (2002). Icosahedral virus particles as addressable nanoscale building blocks. *Angew. Chem.*, 41, pp. 459-462.

Warheit, D. B., Laurence, B. R., Reed, K. L., Roach, D. H., Reynolds, G. A. M. and Webb, T. R. (2004). Comparative pulmonary toxicity assessment of single-wall carbon nanotubes in rats. *Toxicol. Sci.*, 77, pp. 117-125.

Watson, P., Jones, A. T. and Stephens, D. J. (2005a). Intracellular trafficking pathways and drug delivery: Fluorescence imaging of living and fixed cells. *Adv. Drug Deliv. Rev.*, 57, pp. 43-61.

Watson, P., Jones, A. T. and Stephens, D. J. (2005b). Intracellular trafficking pathways and drug delivery: fluorescence imaging of living and fixed cells. *Adv. Drug Deliv. Rev.*, 57, pp. 43-61.

Wattiaux, R., Laurent, N., Wattiaux-De Coninck, S. and Jadot, M. (2000). Endosomes, lysosomes: their implication in gene transfer. *Adv. Drug Deliv. Rev.,* 41, pp. 201-208.

Weyhers, H., Löbenberg, R., Mehnert, W., Souto, E. B., Kreuter, J. and Müller, R. H. (2006). In vivo distribution of 125I-radiolabelled solid lipid nanoparticles. *Pharm. Ind.*, 68, pp. 889-894.

Willis, R. C. (2004). Good Things in small packages. *Modern Drug Discovery*, pp. 30-36.

Wörle-Knirsch, J. M., Pulskamp, K. and Krug, H. F. (2006). Oops they did it again! Carbon nanotubes hoax scientists in viability assays. *Nano Letters*, 6, pp. 1261-1268.

Wu, X., Liu, H., Liu, J., Haley, K. N., Treadway, J. A., Larson, J. P., Ge, N., Peale, F. and Bruchez, M. P. (2003). Immunofluorescent labeling of cancer marker Her2 and other cellular targets with semiconductor quantum dots. *Nat. Biotech.*, 21, pp. 41-46.

Xu, Z. P., Hua Zeng, Q., Lu, G. Q. and Bing Yu, A. (2006). Inorganic nanoparticles as carriers for efficient cellular delivery. *Chem. Engin. Sci.*, 61, pp. 1027-1040.

Yager, P. and Schoen, P. E. (1983). Formation of tubules by a polymerizable surfactant. *Mol. Cryst. Liquid Cryst.* 106, pp. 371-381.

Zauner, W., Farrow, N. A. and Haines, A. M. R. (2001). In vitro uptake of polystyrene microspheres: effect of particle size, cell line and cell density. *J. Control. Rel.*, 71, pp. 39-51.

Zhang, N., Ping, Q., Huang, G., Han, X., Cheng, Y. and Xu, W. (2006). Transport characteristics of wheat germ agglutinin-modified insulin-liposomes and solid lipid nanoparticles in a perfused rat intestinal model. *J. Nanosci. Nanotech.*, 6, pp. 2959-2966.

Chapter 3

TARGETING APPROACHES

Sandrine Cammas-Marion
*UMR 6226 CNRS, Ecole Nationale Supérieure de Chimie de Rennes,
Rennes Cedex, France*

1. Introduction

To ensure a required therapeutic concentration, one has to administrate high and repeated amounts of drug because of its non-selective distribution in the body and of the numerous biological barriers the drug has to cross before reaching the site of action, thus leading to several negative side effects. Therefore, to increase drugs efficacy while decreasing their toxic side effects, it has been imagined to encapsulate the drug in a nanocarrier [Torchilin, 2000; Brannon-Peppas *et al.*, 2004; Minko, 2004; Chellat *et al.*, 2005]. However, drug encapsulation in a nanocarrier, such as liposomes or nanoparticles, is not sufficient to obtain satisfactory therapeutic indices because of a rapid nanocarrier uptake by the reticuloendothelial system as a result of the plasma proteins (opsonins) adsorption on hydrophobic surfaces of the carriers. To overcome this drawback, surface of nanoparticles and liposomes have been coated by a hydrophilic polymer, the poly(ethylene glycol) (PEG), resulting in long term circulation Stealth® nanocarriers [Sapra *et al.*, 2003; Andresen *et al.*, 2005; Chellat *et al.*, 2005]. Such passive targeting can lead, in some cases, to a significant increase of drug accumulation in the disease tissues.

Nevertheless, passive targeting does not always led to effective drug accumulation in a specific tissue or organ. Therefore, to increase the specificity of interaction between nanocarriers and target cells or tissues as well as to increase the amount of drug delivered to the desired site of action, active targeting is needed [Torchilin, 2000; Brannon-Peppas *et al.*, 2004; Minko, 2004; Chellat *et al.*, 2005]. Such active targeting can be obtained by coupling a targeting moiety to the nanocarriers, providing a selective and quantitative accumulation of the nanocarriers, and therefore of the drug, at the target site [Torchilin, 2000; Sapra *et al.*, 2003; Kim *et al.*, 2005; Jaracz *et al.*, 2005]. Several of the specific receptors present at the level of a disease organ, tissue or cell are known. Consequently, adapted targeting moieties can be selected and coupled to nanocarriers in order to design site-specific drug delivery systems. Whatever the way selected for coupling targeting moiety to a nanocarrier, the reaction has to be simple, fast, efficient and reproducible. The coupling method has to yield to stable and non-toxic bonds. Moreover, target recognition and binding efficiency of the coupled molecule have to be maintained. Furthermore, targeted nanocarriers have to be stable enough to present a circulation half life allowing them to reach and interact with their site of action. Finally, both the drug loading efficiency and the drug release profile do not have to be significantly changed by targeting moieties coupling reactions.

This chapter will not be an exhaustive overview of the literature related to the design of site-specific drug carriers, but it will give some examples of targeting approaches for nanotherapeutics. The design of the site-specific nanocarriers, i.e. the different ways of coupling the targeting moieties to the nanocarriers, will be exposed and illustrated with examples giving the effectiveness of coupling and targeting.

2. Coupling of Targeting Moieties on Preformed Nanocarriers

The design of nanocarriers possessing targeting moieties on their surface can be realized by coupling of the selected molecule to the surface of preformed nanocarriers using various methods of the coupling chemistry domain.

Considerable amount of work has been done on the coupling of antibodies on the surface of preformed nanocarriers, often using maleimide groups located on their surface. For example, one possibility is coupling a single-chain Fv fragment (scFv A5) directed against human endoglin to surfaces of preformed liposome loaded with carboxyfluorescein or doxorubicin [Völkel *et al.*, 2004]. The anti-endoglin scFv has been site-specifically coupled to preformed maleimide-containing liposomes via the C-terminal cysteine residues of scFv A5 with a coupling efficiency of about 10-20%. Authors have attributed this low coupling efficiency to a low accessibility of the reactive moieties which were too close to liposome surface. Despite this low contain of targeting moieties, scFv A5 immunoliposomes showed strong and specific binding to endoglin-expressing endothelial cells *in vitro*. Results obtained with rhodamine-labelled scFv A5 immunoliposomes and with scFv A5 immunoliposomes containing carboxyfluoresceine demonstrated that these liposomes were internalized into endothelial cells. *In vitro* cytotoxicity studies have shown an increased cytotoxicity of doxorubicin-loaded scFv A5 immunoliposomes towards endothelial cells. However, *in vivo* pharmacokinetic studies evidenced a rapid clearance from blood circulation of such formulations, probably due to the presence of the functionalized coupling lipid and the use of coupling chemistry. The authors have concluded that the observed limitations can be overcome by reducing number of coupling lipids, the use of Stealth® liposomes or post-insertion method.

In order to increase targeting efficiency of nanocarriers, the coupling of two or more populations of antibodies on liposomes' surfaces has been envisaged [Laginha *et al.*, 2005]. Indeed, the authors have hypothesized that the antigen density can be artificially increased by targeting two or more antigen populations. They have studied binding and uptake of dual-coupled Stealth® immunoliposomes: no statistical differences have been observed between the binding and uptake of liposomes on which a mixture of antibodies was coupled versus that of liposomes bearing one pure antibody type. Laginha *et al.* concluded that it can be possible to observe synergistic interactions with the appropriate choice of dual targeted liposomes but that more experiments will be necessary to evaluate the real efficiency of such systems.

Besides coupling of antibodies to liposome surfaces, such targeting moieties have been also linked to preformed nanoparticles surface. For example, the use of immunotargeted PEG-phosphatidylethanolamine (PEG-PE) micelles for encapsulation and site-specific delivery of meso-tetraphenyl porphine (TPP), a photodynamic therapy reagent has been described [Roby *et al.*, 2006]. The selected antibody, mAb2C5, was bound to micelles by incubation with PEG micelles containing 5 mol% of p-nitrophenylcarbonyl (pNP)-PEG-PE polymer. *In vitro* phototoxicity has been evaluated by studying viability of different cells (LLC, B16, MCF-7, and BT20 cells) upon photoirradiation in the presence of TPP loaded micelles and TPP-loaded mAb2C5 immunomicelles and compared to those obtained for free TPP. Both TPP-loaded micelles and TPP-loaded immunomicelles showed low dark toxicity (more than 95% cell survival even at highest concentrations of TPP). Upon light irradiation, a strong increase in the post-light-irradiation cytotoxicity has been observed for TPP-loaded micelles and TPP-loaded immunomicelles in comparison to free TPP. Roby et al. concluded that such efficacy can be attributed to a better internalization of TPP-loaded formulation by cancer cells.

Recently, amphiphilic block copolymer containing small molecule drug segments and tosylated hexaethylene glycol segments have been described in literature [Bertin *et al.*, 2006]. These copolymers were assembled to form core-shell polymeric nanoparticles. The functional

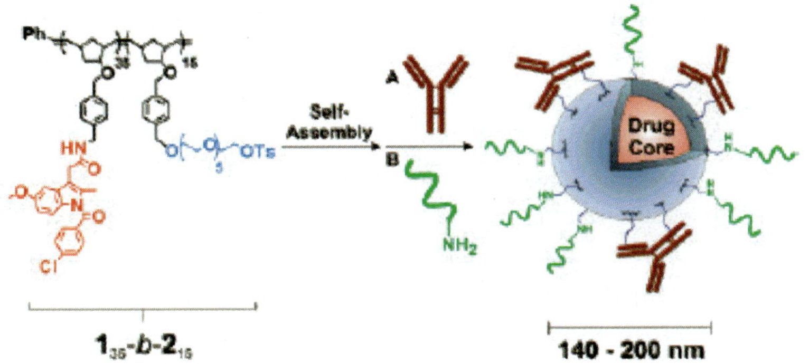

Fig. 1 Preparation of multifunctional polymeric nanoparticles from 135-b-215 [Reprinted with permission from Bertin *et al.*, 2006, Copyright 2006 American Chemical Society].

groups located at nanoparticle surfaces allowed conjugation of single stranded DNA sequences and/or tumor-targeting antibodies (Figure 1).

The immuno-nanoparticles were prepared by incubating aqueous suspensions of tosylated polymeric nanoparticles with anti-HER-2IgY. Immobilization of this antibody was evidenced by transmission electronic microscopy of formulations after immunonanoparticles exposure to gold nanoparticles bearing anti-IgY secondary antibodies. Results showed that surface immobilized anti-HER-2-IgY effectively facilitated the internalization of multifunctional polymeric nanoparticles in SKBR3 human breast carcinoma cells over expressing the HER-2/neu gene.

Besides targeting of nanocarriers by antibodies, others molecules have been coupled on the surface of preformed nanocarriers to achieve active targeting. Li *et al.* [2003] have used transferrin as a targeting moiety for DNA-loaded PEG-poly(cyanoacrylate) nanoparticles. The PEG-poly(cyanoacrylate) nanoparticles loading DNA were prepared by a water-oil-water solvent evaporation technique. Transferrin was then coupled to the nanoparticles using periodate oxidation method. Results of cell association assay have shown that the binding of transferrin-PEG nanoparticles to K562 cells was receptor specific. Authors concluded that transferrin-PEG nanoparticles bearing 1 to 3% of total PEG chains conjugated to transferrin molecules exhibited a higher degree of binding to K562 cells than non-targeted PEG nanoparticles at 4°C. The results are promising but efforts have to be done to increase DNA loading efficiency and *in vitro* and *in vivo* transfection efficiencies have to be studied.

In a comparable study, the *in vitro* and *in vivo* efficiencies of paclitaxel delivery using paclitaxel loaded PEG-poly(cyanoacrylate) nanoparticles were evaluated [Xu *et al.*, 2005]. Transferrin has been coupled as described above leading to 5 to 8% of the total PEG chains linked to transferrin molecules. Pharmacokinetics and biodistribution in mice have evidenced that paclitaxel loaded in targeted and non-targeted nanoparticles was eliminated rather slowly, probably as a result of the PEG presence leading to a decrease in the recognition by mononuclear phagocyte system. Finally, *in vivo* anti-tumor activity studies demonstrated that treatment by paclitaxel loaded transferrin-PEG nanoparticles seem to be quite efficient, since tumor regression was significant with complete tumor regression for five out nine mice and life span of tumor-bearing mice was significantly increased.

Hatakeyama *et al.* [2004] have evaluated the factors governing the *in vivo* tissue uptake of transferrin coupled PEG liposomes. The authors have shown that transferrin-PEG and PEG liposomes have a long circulation with a half-life superior to 6 hours whatever their size. On the other hand, organ distribution of both transferrin-PEG and PEG liposomes was depending on their size. Hatakeyama et al. concluded that a size smaller than 80 nm was an important factor for an efficient tissue targeting (in liver and brain) of transferrin-PEG liposomes based on receptor-mediated endocytosis.

To improve the treatment of cancer by boron neutron-capture therapy, Maruyama *et al.* [2004] have used a borate derivative (BSH) loaded transferrin conjugated PEG liposomes (transferrin-PEG liposomes). Transferrin-PEG liposomes were shown to be effectively receptor specific, bound *in vitro* to Colon 26 cells and internalized by endocytosis. The authors observed that transferrin-PEG liposomes introduced efficiently BSH to Colon 26 cells by a receptor-mediated endocytosis. *In vivo* biodistribution and tumor accumulation of BSH loaded liposomes (diameter of 105-125 nm) confirmed the effectiveness of PEG layer in prolonging circulation time of liposomes after intra venous injection. In addition, transferrin-PEG liposomes allowed a high retention and concentration of ^{10}B in tumor tissues indicating a cellular uptake of such liposomes by transferrin receptor mediated endocytosis. Treatment by BSH loaded transferrin-PEG liposomes associated to thermal neutron irradiation significantly inhibited the cell growth.

In the field of oral administration of peptides and proteins, one of the major problems (rapid degradation of these molecules by proteolytic enzyme in the gut) has been solved by encapsulating peptides and proteins into nanocarriers. However, the quantity of material taken up from the intestine to the circulation was very low. In this context, the lectin mediated transport of nanoparticles across Caco2 and OK cells has been studied [Russel-Jones *et al.*, 1999]. In a first step, the authors have shown that the selected lectins were able to bind to Caco2 cells and to be internalized by these cells. In a second step, they have evidenced that it was possible to stimulate the *in vitro* uptake of nanoparticles using three different lectins (LTB, LUGA and ConA) having different binding specificity. The challenge in the development of such site specific nanocarriers for oral delivery of drugs will consist in manufacturing drug loaded nanoparticles with sufficient surface density of lectin and controlled drug release after their entrance in the circulation.

In the field of oral immunization, Gupta *et al.* [2006] have studied the possibilities to use HBsAg (hepatitis B) loaded lectin-poly(lactic acid-co-glycolic acid) (PLGA) nanoparticles. HBsAg loaded PLGA nanoparticles were prepared by double emulsion method in presence of poly(vinyl alcohol) (PVA) with a loading efficiency of about 54%. Lectin (peanut agglutinin) was covalently coupled to surface hydroxyl group of PVA via glutaraldehyde, in a two steps reaction, with a coupling efficiency of 21%, to confer to nanoparticles a M-cell targeting potential. *In vitro* ligand affinity and activity studies suggested that lectinized nanoparticles retained activity and some sugar specificity as the native lectin. *In vivo* studies remain necessary in order to analyze the interactions of such nanoparticles with the Peyer's patch cells and immune response to hepatitis antigen.

Moreira *et al.* [2001] have studied doxorubicin loaded PEGylated liposomes tagged with a growth factor antagonist (antagonist G) as targeted drug delivery systems for human small cell lung cancer. Cellular association experiments, carried out on H69 and Namalwa (negative control) cell lines, demonstrated that antagonist G targeted liposomes were specifically recognized and internalized in H69 cells through a receptor mediated process leading to intracellular drug accumulation and release to intracellular site of action resulting in cytotoxicity. Pharmacokinetics and biodistribution of liposomes, evaluated in mice, showed that targeted liposomes have a different distribution than the one observed for free antagonist G.

Another class of peptides, a cyclic peptide with the RGD sequence cyclo(-Arg-Gly-Asp-D-Phe-Cys-), has been evaluated for site-specific delivery using doxorubicin loaded carbohydrate based nanoparticles [Bibby et al., 2005]. The accumulation of doxorubicin was high in both the liver and spleen while exposure of doxorubicin to cardiac tissue was low. Following administration, drug accumulated in the tumor, reaching 2.1% of the administered dose at 24 h. A metabolite, suspected to be doxorubicinol or doxorubicinone, was also observed in these tumor samples. This metabolite was not seen in any other tissue and may be attributed to enzymatic activity, a decreased pH or an otherwise altered metabolic state in the tumor. The presence of a metabolite in this tissue alone was indicative of a tumor-specific drug nanoparticle lability, and may present a therapeutic advantage.

Sugars have been also widely used as targeting moieties. Liang *et al.* [2006] have prepared nanoparticles composed of poly(γ-glutamic acid)-b-poly(lactide) processing galactosamide on their surfaces (Figure 2).

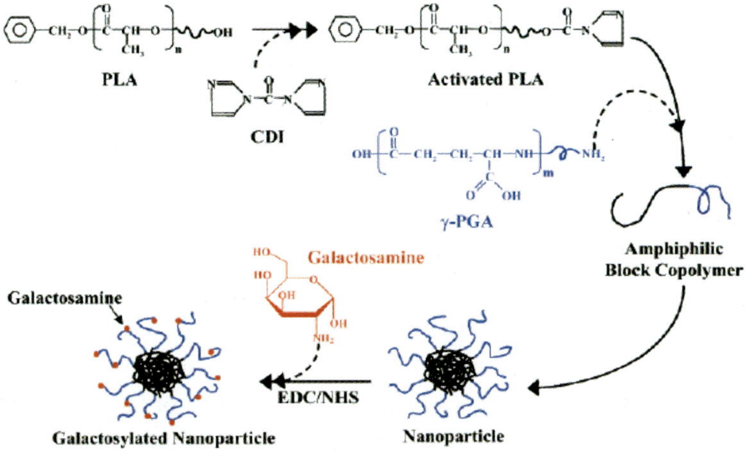

Fig. 2 Schematic illustrations of synthesis of PGA-PLA block copolymers and formation of self-assembled nanoparticles with conjugated galactosamine [Reprinted with permission from Liang *et al.*, 2006, Copyright 2006 American Chemical Society].

Cellular uptake study, using rhodamine-123 loaded PGA-PLA nanoparticle with conjugated galactosomine, indicated that galactosylated nanoparticles had a specific interaction with HepG2 cells via ligand-receptor (ASGP) recognition. Viability of HepG2 cells treated with different paclitaxel formulations showed that the activity in inhibiting the growth of cells by paclitaxel loaded galactosylated PGA-PLA nanoparticles was comparable to that of clinically available paclitaxel (Phyxol®) while paclitaxel loaded PGA-PLA nanoparticles displayed a significantly lower activity. The authors concluded that the galactosylated nanoparticles interacted in a specific manner with HepG2 cells via a ligand-receptor (ASGP) recognition leading to internalization of the drug carrier into HepG2 cells and release of paclitaxel into the cytoplasm. Biodistributions of the prepared nanoparticles in organs of normal mice and hepatoma tumor bearing nude mice showed that galactosylated nanoparticles had specific interactions with liver's parenchymal cells and HepG2 tumor cells via ligand receptor recognition. In addition, anti-tumor efficacy of the prepared nanoparticles on hepatoma tumor bearing nude mice showed that paclitaxel loaded galactosylated PGA-PLA nanoparticles have the higher efficacy in reducing the tumor size [Liang *et al.*, 2006]. The results led the authors to conclude that paclitaxel loaded galactosylated PGA-PLA nanoparticles were mainly accumulated at the tumor site and the liver, in contrast to a non specific accumulation of Phyxol® [Liang *et al.*, 2006].

Among the possible low molecular weight targeting agents, folic acid could be an interesting molecule because this vitamin has receptors frequently overexpressed on the surface of human cancer cells. In this context, Stella *et al.* [2000] studied the design of PEG coated biodegradable nanoparticles conjugated to folic acid for the specific recognition of the soluble form of the folate receptor expressed at the surface of cancer cells. It was found that 14-16 % of the total PEG chains were linked to folic acid molecules. Surface plasmon resonance analysis evidenced that folate nanoparticles were able to effectively recognize the sensorchip-immobilized folate binding protein (FBP). Stella *et al.* explained the greater binding avidity of folate conjugated nanoparticles towards FBP by the fact that these nanoparticles could display a multivalent stronger interaction with FBP.

Coupling of a targeting moiety on surface of preformed nanocarriers has brought a significantly improvement of drug delivery system, at least *in vitro* and with models.

3. Coupling of Targeting Moieties by the Post-Insertion Method

However, because modifications of preformed nanocarriers do not always led to a controlled amount of bound targeting moieties, other ways of coupling have been studied. Recently, a new method for the preparation of liposomes bearing targeting moieties has been developed. This post-insertion technique seems to be relatively simple, leads to an appropriate level of stable ligand incorporation and is not compromising for drug loading efficacy and drug release profile [Iden *et al.*, 2001]. The post-insertion consist in, first, preparing liposomes loaded with the selected drug. In parallel, micelles based on a mixture of PEG-lipid and functionalized PEG-lipid are prepared and the selected targeting moiety is coupled to functionalized PEG-lipid contained in the micelles. Second, the targeting moiety is transferred from micelles to liposomes by incubating both formulations.

In vitro and *in vivo* properties of immunoliposomes made by conventional coupling techniques were compared to those of immunoliposomes prepared by post-insertion method [Iden *et al.*, 2001]. Doxorubicin has been loaded into liposomes before the introduction of the targeting moiety. Lipids and functionalized lipids based micelles were prepared, followed by the coupling of selected antibody (anti-CD19) to the functionalized PEG termini. Then, the antibodies were transferred from micelles to liposomes by incubation at 60°C for one hour. Following transfer, the mixture was purified by chromatography to obtain pure immunoliposomes. *In vivo* therapeutic showed that means survival times for both targeted formulations were higher than those observed for free drug or non-targeted liposomes.

Nielsen *et al.* [2002] have evaluated therapeutic efficacy of anti-ErbB2 immunoliposomes targeted by a phage antibody selected for cellular endocytosis. The authors have shown that F5- immunoliposomes quickly entered the cells via ErbB2-receptor-specific phenomenon with an internalization rate significantly higher than the one observed for un-conjugated scFv. Finally, *in vivo* efficiency of doxorubicin delivered by F5-immunoliposomes using a xenograft model of human ErbB2 over-expressing breast cancer (BT474) has been studied. The authors concluded that tumor regressions for F5-immunoliposomes were significantly superior to those observed for non-targeted liposomes (Doxil®) and far superior to control treatment.

Through these examples, the post insertion technique seems to be simple and to lead to the expected site specific drug nanocarriers. However, number of targeting moieties on carrier surface was not always well defined and a drug leakage was observed during the incubation procedure.

4. Coupling of Targeting Moieties by the Avidin/Biotin Complex

The strong avidin-biotin complex has been used to couple targeting moieties on nanocarrier surfaces with the advantage that no coupling chemistry is normally needed.

Aktas *et al.* [2005] used such a complex to develop chitosan-PEG nanoparticles functionalized with the monoclonal antibody OX26 for brain delivery of caspase inhibitor peptide. A heterobifunctional PEG has been used to covalently coupled biotin followed by covalently coupling of chitosan to lead to a chitosan-PEG-Biotin (CS-PEG-BIO) copolymer. Nanoparticles were prepared by the ionic gelation of pentasodium triphosphate with CS, CS-PEG or CS-PEG-BIO. In parallel, the steptavidin/OX26 conjugate has been prepared and incubated with CS-PEG-BIO nanoparticles to lead to immunonanoparticles. *In vivo* evaluation of the brain uptake of conjugated CS-PEG-BIO nanoparticles has evidenced that the OX26 conjugated to nanoparticles could penetrate into the brain whereas the OX26-free nanoparticles could not. However, further experiments are required to evaluate pharmacological activity of caspase inhibitor peptide loaded nanoparticles.

Vinogradov *et al.* [1999] have studied complexes between oligonucleotides and cationic polymers presenting transferrin on complex surface (Figure 3).

The presence of transferrin can promote uptake of oligonucleotides in the cell and increase transfection. The authors have shown that binding of phosphorothioate oligonucleotides to KBv cell monolayers was 5 to 9 times higher when incorporated into complexes containing avidin than the one observed for avidin-free complex and free phosphorothioate oligonucleotides, respectively. They explained such effect by the binding of avidin to the cell membrane thus stimulating the adsorption-mediated endocytosis of the phosphorothioate oligonucleotide complex containing avidin. Coupling of transferrin to such complexes enhanced the uptake of oligonucleotides into cells. Mdr1 inhibition by oligonucleotides and their complexes, evaluated using multi-drug resistant cells, showed that, despite the high stability of oligonucleotide-PEG-g-PEI complexes, oligonucleotides can be released from the complex and reach its molecular target inside the cell. When transferrin was attached to the avidin-oligonucleotide-complexes, the activity of antisense oligonucleotides incorporated into these complexes was shown to be the most effective in inhibition of the P-glycoprotein functional activity in cancer cells.

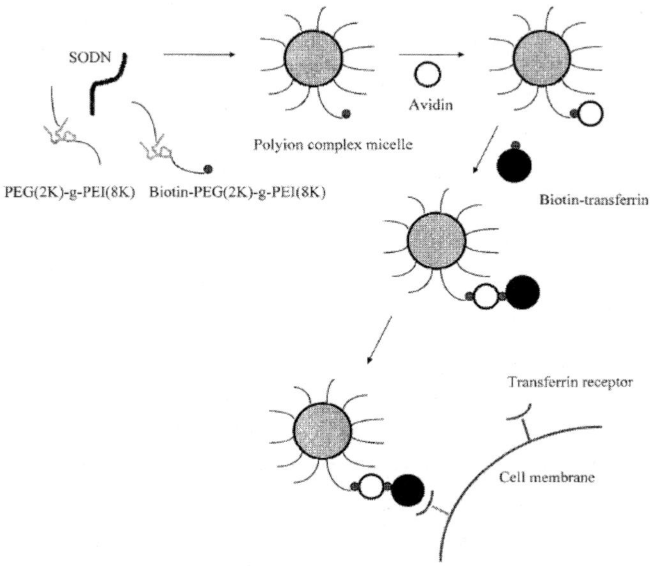

Fig. 3 Schematic illustrating formation of transferrin-modified polyion complex micelle using avidin/biotin construct and its binding to transferrin receptor at the cell surface [Reprinted with permission from Vinogradov *et al.*, 1999, Copyright 1999 American Chemical Society].

This work demonstrated that coupling targeting moieties to a nanocarrier through avidin-biotin interactions is quite simple and can lead to promising results.

5. Coupling of Targeting Moieties before Nanocarriers Formulation

Finally, an efficient method for the introduction of targeting moieties consists in coupling the selected molecule at one end of a lipid or a polymer. Such strategy can be interesting because the coupling chemistry is realized on lipid or polymer, thus allowing easer purification procedures. Moreover, a better control of amount of targeting moieties on nanocarrier surface can be, in theory, reach by introducing a well defined mol% of targeting moiety bearing lipid or polymer in the formulation.

Researchers focused their research on the development of a diblock copolymer micelle system conjugated with epidermal growth factor (EGF) for active targeting of EGF receptor overexpressing cancers [Zeng *et al.*, 2006]. The cellular uptake profile of micelles in EGFR overexpressing MDA-MB-468 cells, followed by fluorescence and microscopy, has shown that targeted micelles were selectively internalized into these cells via an EGF receptor-ligand mediated process. Results have shown that targeted micelles were internalized through a specific endocytotic pathway mediated by the binding of EGF-PEG-b-PVL micelles to cell surface EGFR of MDA-MB-468 cells and were found into perinuclear region and in nucleus. The authors concluded that EGF conjugated copolymer micelles may be used as hydrophobic anticancer drug carriers targeted to EGFR overexpressing cells. Recent studies have used galactose for a selective delivery of various drugs encapsulated in nanoparticles based on different polymers [Cho *et al.*, 2001; Jeong *et al.*, 2005]. They evaluated the use of targetable block copolymer composed of poly(γ-benzyl L-glutamate) and PEG endcapped with galactose moiety (GEG block copolymer) for selective delivery of paclitaxel to liver [Jeong *et al.*, 2005]. Cell cytotoxicity, examined by incubation of P388, SK-Hep01 and HepG2 cells with paclitaxel and paclitaxel loaded GEG nanoparticles, showed that HepG2 cells with asialoglycoprotein receptors (ASGPR) were more sensitive to paclitaxel loaded into GEG nanoparticles than P388 and SK-Hep01 cells without ASGPR. From both cytotoxicity and flow cytometry studies, it was concluded that paclitaxel loaded nanoparticles may be actively delivered to be HepG2 cells with ASGPR through receptor mediated mechanism.

Another approach was to synthesize a novel galactosylated lipid with a good yield by a multi-step synthetic procedure [Wang *et al.*, 2006]. Liposomes were prepared starting from this lipid mixed with others lipids and doxorubicin was entrapped into liposomes by incubation with a loading efficiency of more than 95%. Tissue distribution studies showed that doxorubicin loaded galactosylated liposomes presented a high liver accumulation in comparison to doxorubicin loaded conventional liposomes. Furthermore, the results of intrahepatic distribution and competitive inhibition study evidenced that galactose residues of doxorubicin loaded galactosylated liposomes could be recognized by ASGP on the surface of parenchymal cells leading to high liver accumulation of such targeted liposomes.

A mannosylated cholesterol derivative (Man-C4-Chol) has been prepared by reaction between activated cholesterol and activated mannose derivative [Opanasopit *et al.*, 2002; Hattori *et al.*, 2004]. Liposomes containing this modified lipid were then prepared and muramyl dipeptide (MDP) was loaded into the liposomes. Control liposomes showed a prolonged circulation in plasma and an increasing amount of Man-C4-Chol augmented the uptake of liposomes by the liver. The authors have shown that mannosylated liposomes were recognized by macrophages via the mannose receptor. The *in vivo* biodistribution profiles clearly indicated that the mannosylated liposomes are mainly taken up by the liver non-parenchymal cells. In addition, MDP loaded mannosylated liposomes exhibited excellent activity in preventing liver metastases and significant increase in survival times. Based on the same Man-C4-Chol, the authors prepared mannosylated cationic liposomes for DNA vaccination through targeted gene delivery [Hattori et al., 2004]. It has been shown that mannosylated liposomes/Ovalbumin encoding pDNA complex produced a stronger induction of IL-12, IFN-γ and TNF-α than the un-modified liposome complex. Thus, mannosylated liposomes/pDNA complexes can be efficiently transferred to antigen presenting cells after intravenous administration. However, a higher immunogenicity needs to be attained by modifying the system.

Another group proposed lactose conjugated polyion complex micelles incorporating plasmid DNA as a targetable gene vector system [Wakebayashi *et al.*, 2004]. The authors synthesized a heterobifunctional PEG which was used to polymerize (N,N-dimethylamino) ethyl methacrylate, AMA, leading to acetal-PEG-PAMA copolymer. The lactose moiety was coupled to this copolymer by reacting paminophenyl-β-D-lactopyranoside and aldehyde groups of PEG-PAMA giving access to lactosylated PEG-PAMA. Lactosylated PEG-PAMA/pDNA micelles exhibited higher transfection efficiency against cultured HepG2 cells possessing the asialoglycoprotein receptor in comparison to non targeted micelles due to the contribution of a receptor mediated endocytosis mechanism. To improve their system, the authors [Oishi et al., 2006] introduced a pH responsive polymer into their formulation in order to obtain a targetable and endosome disruptive non viral gene vector (Figure 4). The obtained copolymer spontaneously associated with pDNA to form three layered polyplex micelles. It was shown that the cellular association and internalization of the polyplex micelles occurred mainly through the ASGP receptor mediated process. The transfection efficiency of the lipoplex micelles was significantly improved with an increasing N/P ratio (number of amino groups in the copolymer / number of phosphate groups in pDNA).

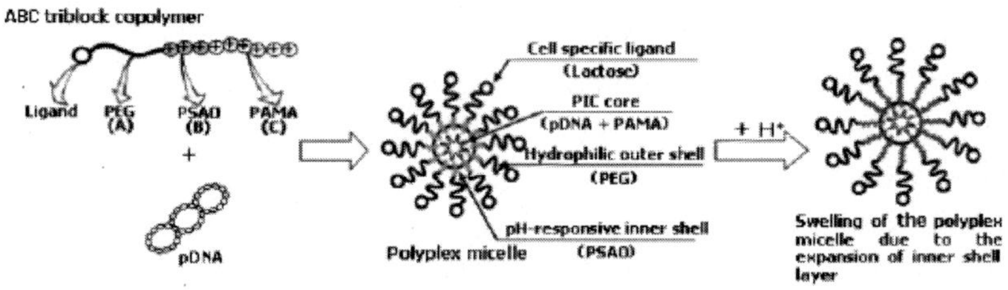

Fig. 4 Schematic illustration of the formation of the three-layered polyplex micelle composed of the ABC triblock copolymer and pDNA [Reprinted with permission from Oishi *et al.*, 2006, Copyright 2006 American Chemical Society].

Confocal microscopy showed that polyplex micelles were localized in the endosomes and/or lysosomes and that some polyplex micelles gradually escaped from the endosomes and/or lysosomes into the cytoplasm in a time dependent manner.

Yoo *et al.* [2004] have prepared biodegradable doxorubicin polymeric micelles having a targeting ability to folate receptor. Doxorubicin was chemically conjugated to the PLGA end of a di-block copolymer composed of PLGA and PEG. Folate was separately conjugated to the end of PEG chain of the PEG-PLGA di-block copolymer. Then, doxorubicin conjugated PLGA-PEG, folate conjugated PLGA-PEG and unprotonated doxorubicin were blended to form self assembled micelles in aqueous solution. *In vivo* animal studies were carried out to examine targeting and anti-tumor effects of doxorubicin/folate micelles using a KB cell xenografted nude mouse model. Doxorubicin/folate micelles suppressed the tumor growth more significantly than free doxorubicin and doxorubicin micelles. This could be attributed to the 'enhanced permeation and retention' (EPR) effect of nano-sized micellar delivery systems. A combined effect of the passive targeting and enhanced cellular uptake would be the main reason for the suppression of tumor growth. Drug–polymer conjugates tend to accumulate at solid tumors by the aforementioned EPR effect, resulting from enhanced permeability of blood vessels on the sites also suggesting that cardiac toxicity of doxorubicin might be significantly reduced by the formulation of doxorubicin micelles.

Stevens *et al.* [2004] synthesized a paclitaxel-cholesterol prodrug and incorporated this prodrug in lipid nanoparticles bearing folate moieties as targeting ligands. The lipid nanoparticle formulations were prepared by solvent dilution followed by diafiltration. Cytotoxicity studies of folate conjugated lipid nanoparticles containing paclitaxel-cholesterol prodrug demonstrated that this prodrug was therapeutically active and that the targeted lipid nanoparticles can effectively target *in vitro* tumor cells over expressing folate receptors. *In vivo* anti tumor efficacy of folate conjugated lipid nanoparticles containing paclitaxel-cholesterol was evaluated in subcutaneous FR(+) M109 tumors in BALB/c mice. Results indicated that mice treated with folate conjugated lipid nanoparticles containing paclitaxel-cholesterol exhibited a reduced tumor growth than mice receiving the non targeted formulations. Moreover, data on survival indicated that folate conjugated lipid nanoparticles containing paclitaxel-cholesterol prodrug were more effective in prolonging the survival of tumor bearing mice than non targeted formulations. The authors concluded that these results reflected a greater therapeutic efficacy due to folate receptor targeting *in vivo*, probably as a result of the higher stability of formulations giving more time to folate conjugated lipid nanoparticles to reach their target site prior to the drug release.

Using the same targeting moiety, folic acid, folate conjugated liposomes for paclitaxel site specific delivery were formulated [Wu *et al.*, 2006]. Unlike their first study, paclitaxel was not coupled to a lipid and folic acid was conjugated to PEG-DSPE via an amide bound. Paclitaxel was incorporated into the liposomes during their formulation with a loading efficiency of about 98 %. The enhancement in cytotoxicity exhibited by paclitaxel loaded folate conjugated liposomes was about 4 fold higher than the one observed with non targeted formulations. This result demonstrated a folate receptor dependence of the cytotoxicity. Pharmacokinetic studies with liposomial formulation of paclitaxel indicated a longer systemic circulation time in comparison to the one observed for the Cremophor® EL micelles. The authors observed that folate conjugated formulations presented a faster clearance than non targeted formulations. However, no data was given concerning *in vivo* biodistribution of folate conjugated liposomes and on the uptake mechanism.

Coupling of targeting moieties to lipid or polymer can be a good solution to introduce the selected targeting molecule on nanocarrier surface in view of the numerous reactions of coupling available.

6. Conclusion

As it is described within this chapter, targeting moieties can be coupled to lipid or polymer constituted the nanotherapeutics before their formulation or introduced in preformed nanotherapeutics using coupling chemistry, biotin-avidin complex or post insertion technique. None of these methods seem to be the ideal one for various reasons such as the difficulties to control amount of targeting moieties on the surface of nanocarriers, leakage of encapsulated drug during coupling procedures, etc.

Nevertheless, since Ehrlich's concept of magic bullet [Ehrlich, 1954], in which the drug was directly bound to a targeting moiety, considerable progresses have been realized in the design of efficient site-specific drug delivery systems. The design and the characterization of such nanotherapeutics become now quite complex in regards to their formulation based on a lot of compounds. Indeed, to reach the most efficient nanocarriers, numerous components have been added to the drug and the targeting moiety of Ehrlich's concept. First, the drug is loaded into carriers constituted by polymers or lipids. To prolong the circulation of the carriers in the body, the hydrophobic surface of nanocarriers has been covered by a hydrophilic polymer, generally the PEG, leading to a passive targeting. In order to deliver the drug to a specific site in the body and to realize an active targeting, targeting moieties have been linked on the surface of the nanotherapeutics. The coupling of targeting moieties has led to significant improvement of nanocarrier efficiency, and in some cases to the successful treatment of the studied disease *in vivo*.

In the research for the ideal nanocarrier presenting the higher efficiency, researchers have also introduced specific bond between the drug and the carrier and/or between the drug and the targeting moiety, bonds which are able to be cleaved in response of a change in the nanocarrier environment (pH, temperature, enzyme, etc.).

References

Aktas, Y., Yemisci, M., Andrieux, K., Gürsoy, R. N., Alonso, M. J., Fernandez-Megia, E., Novoa-Carballal, R., Quinoa, E., Riguera, R., Sargon, M. F., Cüelik, H. H., Demir, A. S., Hıncal, A. A., Dalkara, T., Cüapan, Y. and Couvreur, P. (2005). Development and Brain Delivery of Chitosan-PEG Nanoparticles Functionalized with the Monoclonal Antibody OX26. *Bioconj. Chem.*, 16, pp. 1503-1511.

Andresen, T. L., Jensen, S. S. and Jørgensen, K. (2005). Advanced strategies in liposomal cancer therapy: Problems and prospects of active and tumor specific drug release. *Progr. Lipid Res.*, 44, pp. 68–97.

Bertin P. A., Gibbs J. M., Kwang-Fu Shen C., Thaxton C. S., Russin W. A., Mirkin C. A. and Nguyen S. T. (2006). Multifunctional Polymeric Nanoparticles from Diverse Bioactive Agents. *J. Am. Chem. Soc.*, 128, pp. 4168-4169.

Bibby, D. C., Talmadge, J. E., Dalal, M. K., Kurz, S. G., Chytil, K. M., Barry, S. E., Shand, D. G. and Steiert, M. (2005). Pharmacokinetics and biodistribution of RGD-targeted doxorubicin-loaded nanoparticles in tumor-bearing mice. *Int. J. Pharm.*, 293, pp. 281–290.

Brannon-Peppas, L. and. Blanchette, J. O. (2004). Nanoparticle and targeted systems for cancer therapy. *Adv. Drug Deliv. Rev.*, 56, pp. 1649-1659.

Chellat, F., Merhi, Y., Moreau, A. and Yahia L'H. (2005). Therapeutic potential of nanoparticulate systems for macrophage targeting. *Biomaterials,* 26, pp. 7260-7275.

Cho, C. S., Cho, K. Y., Park, I. K., Kim, S. H., Sasagawa, T., Uchiyama, M. and Akaike, T. (2001). Receptor-mediated delivery of all trans-retinoic acid to hepatocyte using poly(L-lactic acid) nanoparticles coated with galactose-carrying polystyrene. *J. Control. Rel.*, 77, pp. 7–15.

Ehrlich, P. (1954). The partial function of cells. *Int. Arch. Allergy Appl. Immunol.*, 5, pp. 67-86.

Fahmy, T. M., Fong, P. M., Goyal, A. and Saltzman, W. M. (2005). Targeted for drug delivery. *Nanotoday*, pp. 18-26.

Gupta, P. N., Mahor, S., Rawat, A., Khatri, K., Goyal, A. and Vyas, S. P (2006). Lectin anchored stabilized biodegradable nanoparticles for oral immunization1. Development and in vitro evaluation. *Int. J. Pharm.,* 318, pp. 163-173.

Hatakeyama, H., Akita, H., Maruyama, K., Suhara, T. and Harashima, H. (2004). Factors governing the in vivo tissue uptake of transferrin-coupled polyethylene glycol liposomes in vivo. *Int. J. Pharm.,* 281, pp. 25–33.

Hattori, Y., Kawakami, S., Suzuki, S., Yamashita, F. and Hashida, M. (2004) Enhancement of immune responses by DNA vaccination through targeted gene delivery using mannosylated cationic liposome formulations following intravenous administration in mice. *Biochem. Biophys. Res. Comm.*, 317, pp. 992-999.

Iden, D. L. and Allen, T. M. (2001). In vitro and in vivo comparison of immunoliposomes made by conventional coupling techniques with those made by a new post-insertion approach. *Biochim. Biophys. Acta*, 1513, pp. 207-216.

Jaracz, S., Chen, J., Kuznetsova, L. V. and Ojima, I. (2005). Recent advances in tumor-targeting anticancer drug conjugates. *Bioorg. Med. Chem.*, 13, pp. 5043-5054.

Jeong, Y. I., Seo, S. J., Park, I. K., Lee, H. C., Kang, I. C., Akaike, T. and Cho, C. S. (2005). Cellular recognition of paclitaxel-loaded polymeric nanoparticles composed of poly(γ-benzyl L-glutamate) and poly(ethylene glycol) diblock copolymer endcapped with galactose moiety. *Int. J. Pharm.*, 296, pp. 151–161.

Kim, G. J. and Nie, S. (2005). Targeted cancer nanotherapy. *Nanotoday*, pp. 28-33.

Laginha, K., Mumbengegwi1, D. and Allen, T. (2005). Liposomes targeted via two different antibodies: Assay, B-cell binding and cytotoxicity. *Biochim. Biophys. Acta*, 1711, pp. 25-32.

Li, Y., Ogris, M., Wagner, E., Pelisek, J. and Rüffer, M. (2003). Nanoparticles bearing polyethyleneglycol-coupled transferrin as gene carriers: preparation and in vitro evaluation. *Int. J. Pharm.*, 259, pp. 93–101.

Liang, H. F., Chen, C. T., Chen, S. C.,. Kulkarni, A. R., Chiu, Y. L., Chen, M. C. and Sung, H. W. (2006). Paclitaxel-loaded poly(g-glutamic acid)-poly(lactide) nanoparticles as a targeted drug delivery system for the treatment of liver cancer. *Biomaterials*, 27, pp. 2051-2059.

Liang, H. F., Chen, S. C., Chen, M. C., Lee, P. W., Chen, C. T. and Sung, H. W. (2006). Paclitaxel-Loaded Poly(γ-glutamic acid)-poly(lactide) Nanoparticles as a Targeted Drug Delivery System against Cultured HepG2 Cells. *Bioconj. Chem.*, 17, pp. 291-299.

Maruyama K., Ishida O., Kasaoka S., Takizawa T., Utoguchi N., Shinohara A., Chiba M., Kobayashi H., Eriguchi M. a ndYanagie H. Intracellular targeting of sodium mercaptoundecahydrododecaborate (BSH) to solid tumors by transferrin-PEG liposomes, for boron neutron-capture therapy (BNCT) (2004) *J. Control. Rel.*, 98, pp.195–207.

Minko, T. (2004). Drug targeting to the colon with lectins and neoglycoconjugates. *Adv. Drug Deliv. Rev.*, 56, pp. 491-509.

Moreira, J. N., Hansen, C. B., Gaspar, R. and Allen, T. M. ()2001. A growth factor antagonist as a targeting agent for sterically stabilized liposomes in human small cell lung cancer. *Biochim. Biophys. Acta*, 1514, pp. 303-317.

Nielsen, U. B., Kirpotin, D. B., Pickering, E. M., Hong, K., Park, J. W., Shalaby, M. R., Shao, Y., Benz, C. C. and Marks, J. D. (2002). Therapeutic efficacy of anti-ErbB2 immunoliposomes targeted by a phage antibody selected for cellular endocytosis. *Biochim. Biophys. Acta*, 1591, pp. 109– 118.

Oishi, M., Kataoka, K. and Nagasaki, Y. (2006). pH-Responsive Three-Layered PEGylated Polyplex Micelle Based on a Lactosylated ABC Triblock Copolymer as a Targetable and Endosome-Disruptive Nonviral Gene Vector. *Bioconj. Chem.*, 17, pp. 677-688.

Opanasopit, P., Sakai, M., Nishikawa, M., Kawakami, S., Yamashita, F. and Hashida, M. (2002). Inhibition of liver metastasis by targeting of immunomodulators using mannosylated liposome carriers. *J. Control. Rel.*, 80, pp. 283–294.

Roby, A., Erdogan, S. and Torchilin, V. P. (2006). Solubilization of poorly soluble PDT agent, meso-tetraphenylporphin, in plain or immunotargeted PEG-PE micelles results in dramatically improved cancer cell killing in vitro. *Eur. J. Pharm. Biopharm.*, 62, pp. 235-240.

Russell-Jones, G. J., Veitch, H. and Arthur, L. (1999). Lectin-mediated transport of nanoparticles across Caco-2 and OK cells. *Int. J. Pharm.*, 190, pp. 165–174.

Sapra, P. and Allen, T. M. (2003). Ligand-targeted liposomal anticancer drugs. *Progr. Lipid Res.*, 42, pp. 439-462.

Stella, B., Arpicco, S., Peracchia, M. T., Desmaële, D., Hoebeke, J., Renoir, M., D'Angelo, J., Cattel, L. and Couvreur, P. (2000). Design of Folic Acid-Conjugated Nanoparticles for Drug Targeting. *J. Pharm. Sci.*, 89, pp. 1452-1464.

Stevens, P. J., Sekido, M. and Lee, R. J. (2004). A Folate Receptor–Targeted Lipid Nanoparticle Formulation for a Lipophilic Paclitaxel Prodrug. *Pharm. Res.*, 21, pp. 2153-2157.

Torchilin, V. P. (2000). Drug targeting. *Eur. J. Pharm. Sci.*, 11, pp. S81-S91.

Vinogradov, S., Batrakova, E., Li, S. and Kabanov, A. (1999). Polyion Complex Micelles with Protein-Modified Corona for Receptor-Mediated Delivery of Oligonucleotides into Cells. *Bioconj. Chem.*, 10, pp. 851-860.

Völkel, T., Hölig, P., Merdan, T., Müller, R. and Kontermann, R. E. (2004). Targeting of immunoliposomes to endothelial cells using a single-chain Fv fragment directed against human endoglin (CD105). *Biochim. Biophys. Acta*, 1663, pp. 158-166.

Wakebayashi, D., Nishiyama, N., Yamasaki, Y., Itaka, K., Kanayama, N., Harada, A., Nagasaki, Y. and Kataoka, K. (2004). Lactose-conjugated polyion complex micelles incorporating plasmid DNA as a targetable gene vector system: their preparation and gene transfecting efficiency against cultured HepG2 cells. *J. Control. Rel.*, 95, pp. 653-664.

Wang, S. N., Deng, Y. H., Xu, H., Wu, H. B., Qiu, Y. K. and Chen, D. W. (2006). Synthesis of a novel galactosylated lipid and its application to the hepatocyte-selective targeting of liposomal doxorubicin. *Eur. J. Pharm. Biopharm.*, 62, pp. 32-38.

Wu, J., Liu, Q. and Lee, R. J. (2006). A folate receptor-targeted liposomal formulation for paclitaxel. *Int. J. Pharm.*, 316, pp. 148-153.

Xu, Z., Gu, W., Huang, J., Sui, H., Zhou, Z., Yang, Y., Yan, Z. and Li, Y. (2005). In vitro and in vivo evaluation of actively targetable nanoparticles for paclitaxel delivery. *Int. J. Pharm.*, 288, pp. 361–368.

Yoo, H. S. and Park, T. G. (2004). Folate receptor targeted biodegradable polymeric doxorubicin micelles. *J. Control. Rel.*, 96, pp. 273–283.

Zeng, F., Lee, H. and Allen, C. (2006). Epidermal Growth Factor-Conjugated Poly(ethylene glycol)-*block*-Poly(δ-valerolactone) Copolymer Micelles for Targeted Delivery of Chemotherapeutics. *Bioconj. Chem.*, 17, pp. 399-409.

Part II

DISEASE-RELATED APPROACHES BY NANOTHERAPEUTICS

Chapter 4

NANOSCALE CANCER THERAPEUTICS

Yann Pellequer and Alf Lamprecht

Laboratory of Pharmaceutical Engineering,
Faculty of Medicine and Pharmacy, University of Franche-Comté,
Besançon (France)

1. Introduction

Although many new drugs are synthesised to treat cancer diseases, the clinical potential of such structures is subject of certain therapeutic and toxicological limitations, mainly depending on the physicochemical properties of the drug. Among them, the most important are the simple barrier effect of membranes on cellular or tissue level, pathophysiological drug resistance mechanisms by the cells, and biodistribution behaviour of the drug. At the tumor level, the drug transport being governed also by the physicochemical properties of the interstitium (composition, structure, charge) and of the molecule properties (size, configuration, charge, hydrophobicity) [Jain, 2007].

Because the body distribution of an anticancer drug is essentially based on its physicochemical properties, hydrophilicity, polarity, electrostatic charge, which are not necessarily fitting the characteristics of the diseased area, high concentrations of drug can be distributed towards tissues other than the target. Subsequently, higher drug doses are necessary and toxicity is triggered by the increased drug penetration into healthy organs and tissues, which is one main limiting factor of tumor therapy [Brigger *et al.*, 2002]. At the cellular level, the resistance of tumors to therapeutic intervention may be caused by alterations in the biochemistry of malignant cells including altered apoptosis regulation, or transport-based mechanisms, such as P-glycoprotein efflux system, responsible for the multidrug resistance or the multidrug-resistance-associated protein [Krishna *et al.*, 2000]. In consequence, conventional chemotherapeutics are often inadequately delivered to the tumor target tissue.

The targeting strategies to solid tumors are similar independently from the nanocarrier type used for the formulation. A passive targeting which is based on a plain polymeric nanoparticle design was proposed in the experimental therapy of hepatic cancer types.

At tissue level, upon intravenous injection, colloids are opsonised and rapidly cleared from the blood stream by the normal reticuloendothelial defense mechanism, irrespective of particle composition [Kreuter *et al.*, 1979]. Thus, the liver accumulates essential quantities of nanoparticles, liposomes, etc., conditioning their rapid first-phase disappearance from the blood, followed by degradation and excretion. This biodistribution considered to be beneficial for the therapy of tumors located close by the mononuclear phagocyte system, e.g. hepatocarcinoma or hepatic metastasis arising from digestive tract or gynecological cancers. As the mononuclear phagocyte system is not exclusively located in the liver other therapeutic strategies are feasible such as the therapy of bronchopulmonary tumors - primitive tumors or metastasis - including nonsmall cells tumor and small cells tumors, myeloma, and leukaemia which will be addressed later in this chapter.

In this context it appeared to be promising to consider therapeutic strategies based on drug concept using nanotechnological approaches (liposomes, nanoparticles, polymerized micelles, etc.) to overcome the before mentioned drawbacks. Drug delivery to tumors at the cellular or tissue level permits to improve the specificity of the carried anticancer molecules and, thus its specificity towards the targeted tumor. Strategies for developing new efficient targeted nanoformulations of anticancer compounds is resulting essentially from the combined knowledge of cancer physiopathology features and the adapted design of nanotechnological drug delivery systems.

The drug delivery strategy behind the above mentioned so-called stealth® carriers is that they are not taken up by to macrophages which significantly reducing their rapid blood clearance and recognition by the mononuclear phagocytose system. A major breakthrough in the liposome field consisted in the use of phospholipids substituted with PEG chains [Papahadjopoulos *et al.*, 1991] providing a hydrophilic particle surface, which refrains plasma proteins from adsorption. These "sterically stabilized" liposomes have circulating half-lives of up to 45 hours, as opposed to a few hours or even minutes for conventional liposomes. They have been shown to function as reservoir systems and can penetrate into sites such as solid tumors [Gabizon *et al.*, 1994]. It is believed that these nanosystems need to be small enough and to circulate for a sufficient period of time to extravasate selectively through the small defects of the fenestrated and leaky vasculature that generally characterize tumor vessels [Dvorak *et al.*, 1988; Moghimi *et al.*, 2001].

The already before mentioned enhanced permeability and retention effect (EPR) results in intratumoral drug accumulation, which is even higher than that observed in plasma and other tissues [Noguchi *et al.*, 1998; Maeda *et al.*, 2000]. A similar strategy has been applied to nanoparticles.

The overall literature review confirms that, although an enormous number of cancer types are known the mechanistic approach of nanotechnological drug delivery relies on a few key strategies which are repeatingly encountered in the different therapeutic approaches. Since the development of cancer therapeutics is the major field of research, this field merits in principal an entire book on its own. We will try to give here an overview on principal therapeutic approaches for the different cancer types.

2. Lung Cancer

In the therapy of lung cancer there exist principally two major possibilities to the delivery the anticancer drug, intravenous administration and pulmonary delivery. Due to local toxicity but also security aspects (aerosol distribution – risk for the hospital personnel) the pulmonary pathway has been rather limited attention up to now.

Caelyx® was tested in locally advanced or metastatic non-small cell lung cancer patients, progressed after platinum-based first-line chemotherapy [Numico *et al.*, 2002]. Seventeen patients were enrolled in the study and were considered eligible for evaluation of toxicity and response. Stomatitis, hand-foot syndrome, and asthenia were the most common toxicities and affected approximately half of the treated patients. One confirmed partial response was observed (5.8%); five patients (29.4%) had stable disease (including one minor response) and nine (52.9%) had disease progression. Median time to progression was 9.5 weeks, median survival 18.6 weeks. Caelyx® at the doses employed in this study can be safely administered, but show only limited therapeutic effects

Another report determined the efficiency of Caelyx® in combination with cyclophosphamide and vincristine for previously treated patients with relapsed or refractory small-cell lung cancer [Leighl *et al.*, 2003]. Antitumor activity was seen at all dose levels and this combination is well tolerated.

All-trans retinoic acid was trapped into cationic liposomes in order to inhibit tumor cell growth in established metastatic lung tumors by delivery to the pulmonary tumor site after intravenous injection in mice [Suzuki *et al.*, 2006]. After intravenous injection, the highest lung accumulation of the drug was observed by the cationic liposomal formulation and reduced the number of tumor nodules compared with controls of drug solution or anionic liposomes.

It has been repeatedly demonstrated that long-circulating PEG-grafted liposomes display an increased accumulation in solid tumors via the EPR effect [Gabizon, 1992; Papahadjopoulos *et al.*, 1991]. As with long-circulating liposomes, micelles formed by PEG_{750}–PE, PEG_{2000}–PE, and PEG_{5000}–PE accumulate efficiently in tumors [Lukyanov *et al.*, 2002]. Micelle formulations from all three PEG–PE conjugates studied demonstrated much higher accumulation in tumors compared to non-target tissue (muscle) in experimental Lewis lung carcinoma in mice.

Also in the lung cancer therapy several innovative therapeutic approaches based on targeted nanocarriers have been described in the literature.

Targeting of immunoliposomes to pulmonary endothelial cells of the lungs was found possible using the IgG monoclonal antibody (34A) directed toward the glycoprotein receptor pp120 [Maruyama *et al.*, 1999]. Antagonist G-targeted liposomes increased the targeting of doxorubicin toward the human small-cell lung cancer H69 cell line as revealed by the increased cellular uptake of the targeted liposomes compared with the nontargeted liposomes [Moreira *et al.*, 2002].

In order to increase the selectivity of the drug delivery system immune targeted micelles were proposed recently for a paclitaxel treatment of experimental mice bearing Lewis lung carcinoma [Torchilin *et al.*, 2003]. Immunomicelles with attached antitumor mAb 2C5 effectively recognized and bound various cancer cells in vitro and showed an increased accumulation in experimental tumors in mice when compared with nontargeted micelles. Moreover, compared to non-targeted micelles the new carriers exhibited an enhanced tumor growth inhibition.

Protamine, an arginine-rich peptide, may be used to condense DNA before complexation or encapsulation with the above-mentioned cationic lipids. This mixture has been used for the delivery of the tumor suppressor genes Rb or E1A [Ueno *et al.*, 2002; Nikitin *et al.*, 1999]. This resulted in tumor cell apoptosis, reduction of tumor growth, increased life span in experimental human xenograft models, and in spontaneous multiple neuroendocrine neoplasia and lung metastasis in Rb +/− mice [Nikitin *et al.*, 1999]. The diverse approaches in cancer gene therapy are extensively reviewed elsewhere [El-Aneed, 2004].

Although inhaled drug formulations in lung cancer therapy have already entered clinical trials, such approaches rely on micrometric systems and application of nanocarriers in this context remains experimental for the moment [Sandler *et al.*, 2007]. The inhalation delivery of 5-fluorouracil in lipid-coated nanoparticles to hamsters was evaluated to determine the feasibility for use in lung cancer chemotherapy [Hitzmann *et al.*, 2006]. Within 24 hours, more than 99% of the particles were cleared from the respiratory tract and from an eight-compartment pharmacokinetic model analysis, effective local targeting as well as sustained efficacious concentrations of 5-fluorouracil in the expected tumor sites were demonstrated. However, therapeutic efficiency remains to be determined in future experiments.

3. Hepatic Cancers

When anthracycline antitumor agents were encapsulated into liposomes, they show reduced cardiac as well as gastrointestinal toxicities because the major part of the injected dose is sequestered into the mononuclear phagocyte system, which provides lower peak plasma levels while maintaining similar total body exposure than the free drug counterparts [Gabizon *et al.*, 1989]. It was suggested that after the drug-loaded liposomes are captured by the Kupffer cells of the liver, the liposome matrix becomes leaky, and the drug (and its active metabolites) may be released and distributed in free form to the tumor. The therapeutic index is improved because the anthracycline's antitumor efficacy is maintained, whereas acute and chronic toxicities are substantially reduced.

The efficiency of doxorubicin targeted with the aid of poly(alkylcyanoacrylate) nanoparticles has been demonstrated in a murine hepatic metastases model [Chiannilkulchai *et al.*, 1990]. Besides, with such nanoparticles loaded with doxorubicin as well as with other nanocarriers, interestingly reduced impact by the multidrug resistance was observed. This was probably due to the strong adsorption of nanoparticles onto the cell surface induces a microgradient of drug concentration at the membrane, which, in turn, increases the intracellular diffusion of doxorubicin, thus overflowing the P-glycoprotein detoxification capacity [Verdière *et al.*, 1997, Lamprecht *et al.*, 2006]. Another mechanism discussed in this context is the high drug load delivered into the cell by the nanocarrier taken up by a pinocytotic mechanism acting as some kind of "trojan horse" [Garcion *et al.*, 2006]. Here, the drug is delivered intracellularly by the nanocarrier far from potential influences such as P-glycoproteins.

The higher cytotoxicity of doxorubicin when loaded onto poly(isohexylcyanoacrylate) nanoparticles has been shown recently on the X/*myc* transgenic mouse model of hepatocellular carcinoma, which mimics several steps of human hepatocarcinogenesis. In this study, doxorubicin-loaded poly(isohexylcyanoacrylate) nanoparticle-induced apoptosis was specific and restricted to hepatocellular carcinoma tumors because it did not enhance the apoptosis rate of noncancer hepatocytes in peritumor areas [Barraud *et al.*, 2005]. Less promising results based on the same therapeutic strategy were obtained in a recent study where a HDCC hepatic tumor model was induced in rats and the benefit of passive targeting by lipid nanoparticles was not significantly different from the free drug control [Lacoeuille *et al.*, 2007].

Results leave open whether there is an essential clinical benefit by the nanotechnological liver targeting. Besides, there is an unambiguous risk of liver toxicity as an adverse effect by these systems as high anticancer drug loads come also in contact with healthy liver tissue.

A lot of efforts have been devoted in achieving "active targeting" to deliver drugs to the right cells, based on molecular recognition processes. Specific antibodies or ligand targeting proteins expressed on cancer cell membranes or endothelial cells lining the newly generated blood vessels into the tumor are among the possible options to perform the active targeting of nanotechnologies toward tumoral sites. Examples of relevant targets include galactolipids that bind to the asialoglycoprotein receptor of the human hepatoma HepG2 cells. An optimal coating of 10–30 antibody molecules per liposome seems to allow the combination of an efficient delivery with a limited uptake by the MPS [Maruyama *et al.*, 1999]. Beside macromolecular targeting approaches also small molecules such as mannose can be considered to design targeted devices. It was used to target immunomodulators to liver metastasis with mannosylated liposomes [Sudimack *et al.*, 2002]. Especially the last cited examples are all very advanced strategies which work in animal models however, did not reach the market yet.

4. Renal Cancer

The benefit of nanocarriers in the therapy of renal cancer appears to be very limited. A phase II trial of liposomal encapsulated doxorubicin in patients with advanced renal cell carcinoma showed that none of the fourteen evaluable patients achieved a complete or partial response. No cardiac toxicity was evident, however 79% of patients experienced grade III or IV neutropenia [Law *et al.*, 1994].

In a phase II study of Caelyx® on patients with refractory renal cell cancer [Skubitz, 2002]. Toxicities were mild and similar to previous reports but dose reduction per the study protocol, which was designed to control the skin and mucosal toxicities, was common. Again no definite cardiac toxicity was observed. However, no objective responses to treatment were observed in the patients. This study did not demonstrate activity of pegylated-liposomal doxorubicin in renal cell cancer, although it can be given with mild toxicity.

Other nanotechnological systems were tested, e.g. immunoliposomes for selective targeting [Singh *et al.*, 1991], but these approaches were apparently not continued into clinics. In consequence, the described existing drug delivery systems are adapted to other, potentially more efficient, drugs [Stathopoulos *et al.*, 2005].

5. Ovarian Cancers

Current therapeutic strategies in these malignancies include the use of moderately effective initial regimens that are usually accepted by patients. Tolerability considerations are especially important in the development of palliative regimens: retreatment for persistent or hormone-resistant disease must include quality-of-life analyses [Muggia, 1997]. In early clinical studies in patients with refractory ovarian cancer, polyethyleneglycol coated doxorubicin-containing liposomes has produced high response rates (26%) and gratifyingly long response durations (8 to 21 months after onset of therapy). Information from these same clinical studies confirms the marked reduction in several toxicities associated with free doxorubicin, including nausea and vomiting, myelosuppression and cardiotoxicity. Alopecia is also markedly diminished. On the other hand, mucosal and skin toxicities appear to be more common.

Doxil®/Caelyx®, has been investigated in various cancer types including breast cancers, ovarian cancer, non-Hodgkin's lymphoma, nonsmall cell lung cancer, etc. In the USA, Doxil® was approved by the Food and Drug Administration for the treatment of metastatic ovarian cancer in patients with diseases refractory to both paclitaxel- and platinum-based chemotherapy regimens, and it may be considered as a drug of choice for patients with advanced ovarian cancer for whom first-line chemotherapy has failed [Rose, 2005]. Indeed, pegylated doxorubicin liposomes have demonstrated a significant pharmacological efficacy in the treatment of recurrent or relapsed ovarian cancers in several clinical trials [Gordon *et al.*, 2001; Rose, 2005]. Because the long circulating liposomes promote extravasation of the drug, new toxicities may emerge, the most common being the hand–foot syndrome [Lyass *et al.*, 2000].

Other clinical studies warn of the unique toxicity profile and a delay of doses for subsequent cycles was required with multiple dosing [Fujisaka *et al.*, 2006]. Therapeutic efficiency may vary, as in this case objective response was observed in one out of 15 patients and the normalization of tumor marker values in another.

Besides, several alternative approaches which are nearly all in the experimental stage for the moment are found in the literature.

Liposomes have been used as a method to overcome some delivery issues and, in combination with hyperthermia, have been shown to increase drug delivery to tumors [Kong *et al.*, 2000]. At 34°C, no liposomes were able to extravasate into the tumor interstitium however, hyperthermia enabled liposome extravasation. The magnitude of hyperthermia-induced extravasation was inversely proportional to particle size. At 42°C, the pore cutoff size was increased to >400 nm and 100-nm liposomes experienced the largest relative increase in extravasation from tumor vasculature. Hyperthermia did not enable extravasation of 100-nm liposomes from normal vasculature, potentially allowing for tumor-specific delivery.

When paclitaxel loaded nanoparticles were administered intraperitoneally to carcinoma xenograft bearing Fisher344 rats, they significantly reduced tumor weight and ascites volume, and induced apoptosis of tumor cells [Lu *et al.*, 2007]. Moreover, paclitaxel concentrations of pelvic lymph nodes in nanoparticle treated animals werer 20-fold higher than that of animals treated with the standard formulation. This interesting approach is apparently based on a lymphatic targeting, further mechanisms need to be clarified.

Another strategy proposed is the use of ultrasound enhancing the nanocarrier uptake towards the tumor site [Gao *et al.*, 2004]. Although the approach was widely described for other cancer types, only one study is found for ovarian cancer. By using drug free polymeric micelles, the authors found a highly increased deposition of their carriers inside the tumor after intravenous or intraperitoneal administration suggesting such carriers as a promising approach.

6. Breast Cancer

Much preclinical and clinical research focused on the use of nanocarriers in the treatment of breast cancer. Anthracyclines are some of the most active agents in the treatment of breast cancer, and are widely used in all stages of disease. However, cardiac toxic effects are also here the limiting factor of these agents. Trastuzumab, a monoclonal antibody that targets ERBB2, has improved treatment of this aggressive form of breast cancer [Slamon *et al.*, 2001; Romond *et al.*, 2005]; however, its use is limited by a risk of cardiac toxic effects, which occur almost exclusively in patients previously treated with anthracyclines.

Liposomal anthracycline formulations were developed to improve the therapeutic index of conventional anthracyclines, while maintaining their widespread antitumor activity. Liposomal doxorubicin has been compared with conventional doxorubicin in first-line treatment of patients with metastatic breast cancer [Batist *et al.*, 2001]. Efficacy did not differ significantly between the two groups (response rate 43% vs 43% and median survival 19 vs 16 months). Overall, patients assigned liposomal doxorubicin were 80% less likely to develop cardiac toxic effects than were those assigned conventional doxorubicin.

Pegylated liposomal doxorubicin was compared with conventional doxorubicin in patients with previously untreated metastatic breast cancer [O'Brien *et al.*, 2004]. Both agents had similar efficacy, with response rates of 33 and 38%, respectively. The risk of cardiac toxic effects was significantly higher in patients assigned doxorubicin than in those assigned pegylated liposomal doxorubicin. Neutropenia and gastrointestinal toxic effects were reported more commonly with doxorubicin, whereas palmar-plantar erythrodysaesthesia was more common with pegylated liposomal doxorubicin.

The taxanes paclitaxel and docetaxel are some of the most important agents in the treatment of solid tumors, and are also used in all stages of breast cancer. Both drugs are highly hydrophobic, and have to be delivered in synthetic vehicles (polyethylated castor oil for paclitaxel and polysorbate-ethanol for docetaxel). The toxic effects associated with both taxanes are increasingly recognised to be cause by these synthetic vehicles, and not the agents themselves [Gelderblom *et al.*, 2001]. Several new formulations of these agents have been developed in an attempt to decrease the toxic effects associated with the taxanes. A nanoparticle with a core containing paclitaxel surrounded by albumin has shown efficacy in breast cancer. Preclinical studies showed that such nanoparticles resulted in improved tumor penetration compared with conventional paclitaxel. In addition, it resulted in a higher plasma clearance and larger volume of distribution than did paclitaxel, consistent with a lack of sequestration by castor-oil micelles [Sparreboom *et al.*, 2005].

A phase II trial in patients with metastatic breast cancer showed a response of 48% to albumin paclitaxel nanoparticles at a dose of 300 mg/m2 every 3 weeks [Ibrahim *et al.*, 2005]. Overall response was significantly higher in patients allocated albumin paclitaxel nanoparticles compared with those allocated the conventional formulation, irrespective of line of therapy. Although overall survival was not significantly different in the patients, patients in the second-line setting had a significantly higher survival with paclitaxel nanocarriers at 56 weeks compared with conventional paclitaxel at 47 weeks. This nanoparticle formulation of paclitaxel offers advantages over castor-oil-based paclitaxel, with an overall decrease in toxic effects and enhanced efficacy.

About two-thirds of breast cancers express hormone receptors, of which about 50% benefit from endocrine therapy. Tamoxifen remains widely used in all stages of breast cancer, in both premenopausal and postmenopausal women. It undergoes substantial metabolism, and an inability to get active drug into breast tumors might hinder its effectiveness. Tamoxifen-loaded, polymeric nanoparticles were proposed to increase tumor penetration [Shenoy *et al.*, 2005]. By use of a human breast-cancer xenograft model, they showed a significant increase in the level of tumor accumulation of tamoxifen in mice given the loaded nanoparticles, compared with those given an intravenous formulation.

In another study a parenteral delivery system for the administration of the highly promising pure antiestrogen RU 58668 was developed [Ameller *et al.*, 2003]. Two types of nanoparticles made of biodegradable copolymers and coated with PEG chains were compared. Coating with PEG chains prolonged the antiestrogenic potency of drug, as shown by a prolonged antiuterotrophic activity of encapsulated drug into PEG-PLA nanoparticles, as compared to that of conventional nonpegylated nanoparticles. In mice bearing MCF-7 estrogen-dependent tumors, free drug injected at 4.3 mg/kg/week by i.v. route slightly decreased the estradiol-promoted (0.5 mg/kg/week) tumor growth while drug loaded PEG-PLA nanoparticles injected at the same dose strongly reduced it.

Major strategies in breast-cancer gene therapy include transfer of tumor-suppressor genes, enhancement of immunological response, transfer of suicide genes, and bone-marrow protection by use of drug-resistance genes. Breast-cancer genome abnormalities for which gene therapy could be potentially useful include amplification or mutation of multiple genes, including ERBB2, P53, MYC, and cyclin D1 [Osborne *et al.*, 2004].

Nanoparticle-based DNA and RNA delivery systems offer several potential advantages for gene delivery to various human tumors, including breast cancer. A DNA plasmid can be coupled with cationic and neutral lipids to form lipid–nucleic-acid nanoparticles [Hayes *et al.*, 2006]. In addition, conjugation of a polyethylene glycol molecule to the surface of the nanoparticle with targeted antibody increases gene delivery into tumor cells. Preclinical studies have shown that adenovirus type 5 E1A is associated with antitumor activities by transcriptional repression of ERBB2 [Yan *et al.*, 1991]. Another group showed antiproliferative activity of wild-type P53-loaded nanoparticles in a breast-cancer cell line [Prabha *et al.*, 2004].

Transfection of tumor cells with small-interfering RNA (siRNA) is a rapidly growing gene-silencing technology with great potential for clinical application. Inhibition of breast-cancer oncogenes results in induction of apoptosis and an increase of chemotherapy sensitivity in breast-cancer cells [Choudhury *et al.*, 2004]. Stability and cellular uptake of siRNA can be greatly improved by adsorption onto nanoparticles [Schwab *et al.*, 1994]. Nanoparticle–siRNA complexes directed to Ras matrix RNA selectively inhibited the proliferation of breast-cancer cells and markedly inhibited Ha-ras-dependent tumor growth in nude mice after injection under the skin. Despite this early stage of development, nanoparticle-based delivery systems have already shown significant benefits for targeted gene delivery, and indicate great potential for clinical use in breast-cancer therapy.

7. Prostate Cancer

Doxil®/Caelyx® has anti-tumor activity against Kaposi's sarcoma and other solid tumors with mild myelosuppression, minimal hair loss and a low risk of cardiotoxicity. Non-liposomal doxorubicin has modest activity in hormone-refractory prostate cancer with considerable toxicity. A pilot study of Doxil was conducted in 15 patients [Hubert *et al.*, 2000]. Doxil was administered intravenously using two regimes of equal dose intensity, either 45 mg/m2 every 3 weeks or 60 mg/m2 every 4 weeks. Three patients responded to treatment (based on objective response in one patient and reduction of PSA level greater than 50% in the other two) and two patients had stable disease, all of them receiving 60 mg/m2. Doxil at 60 mg/m2 every 4 weeks appears to be active against hormone-refractory prostate cancer, but severe mucocutaneous toxicities prevented further investigation of this regime.

On the basis of doxorubicin's liposomal encapsulation demonstrated clinical efficacy against hormone-refractory prostate carcinoma, a prospective, randomized Phase II clinical trial was conducted to evaluate the feasibility, toxicity, and therapeutic efficacy associated with the pegylated form [Heidenreich *et al.*, 2004]. Pegylated liposomal doxorubicin yielded a noteworthy objective palliative response rate and a mean survival of 13 months for patients with symptomatic hormone-refractory prostate cancer. The dosage tested in the current study appeared to be adapted to future trials with pegylated liposomal doxorubicin-containing combination regimens.

8. Gastric Cancer

The delivery of adriamycin to the regional lymph nodes of the stomach following the gastric submucosal injection of liposomal adriamycin was investigated in 34 gastric carcinoma patients, as well as following intravenous administration of free adriamycin in another 18 patients [Akamo *et al.*, 1994]. Prior to radical gastrectomy, liposomal adriamycin was endoscopically injected into the gastric submucosa adjacent to the primary tumor via a needle-tipped catheter. After liposomal adriamycin injection, the adriamycin concentration in the primary and secondary drainage lymph nodes was higher than in the other regional lymph nodes. Thus, the regional nodes more susceptible to metastasis showed higher levels of adriamycin. In contrast, the intravenous administration of free adriamycin produced a similar and far lower adriamycin concentration in all the nodes. Such preoperative adjuvant chemotherapy targeting the regional lymph nodes may be useful for preventing the lymph node recurrence of gastric carcinoma.

Lipophilic photosensitizers hold potential for cancer photodynamic therapy. A study reported to develop a novel photosensitive stealth liposomes which incorporating a lipophilic photosensitizer into its lipid bilayer and to examine its photoactivity. In gastric cancer cell lines, LC80 values of liposomes was a maximum of 53 times as low as that of Ce6 sodium salt. Liposomes completely destroyed all tumors in animal models and tumor recurrence levels were minimal (1.5±0.9%). photosensitive stealth liposomes achieved greater photodynamic effects in gastric cancer cell lines and in murine models than Ce6-Na holding promise for photodynamic therapy for gastric cancer.

9. Colon Cancer

The antitumor effect of doxorubicin conjugated to a biodegradable dendrimer was evaluated in mice bearing C-26 colon carcinomas [Lee *et al.*, 2006]. In culture, dendrimer-doxorubicin was 10-fold less toxic than free doxorubicin toward C-26 colon carcinoma cells after exposure for 72 h. Upon i.v. administration to BALB/c mice with s.c. C-26 tumors, dendrimer-doxorubicin was eliminated from the serum with a half-life of 16 +/- 1 h, and its tumor uptake was ninefold higher than i.v. administered free doxorubicin at 48 h. In efficacy studies performed with BALB/c mice bearing s.c. C-26 tumors, a single i.v. injection of dendrimer-doxorubicin at 20 mg/kg doxorubicin equivalents 8 days after tumor implantation caused complete tumor regression and 100% survival of the mice over the 60-day experiment. No cures were achieved in tumor-implanted mice treated with free doxorubicin at its maximum tolerated dose (6 mg/kg), drug-free dendrimer, or dendrimer-doxorubicin in which the doxorubicin was attached by means of a stable carbamate bond. The antitumor effect of dendrimer-doxorubicin was similar to that of an equimolar dose of liposomal doxorubicin (Doxil®). The remarkable antitumor activity of dendrimer-doxorubicin results from the ability of the dendrimer to favorably modulate the pharmacokinetics of attached doxorubicin.

In studies comparing stealth® liposomal cisplatin (SPI-77) and cisplatin tumor disposition in murine colon tumor xenografts, the platinum (Pt) exposure was four-fold higher and prolonged after SPI-77 compared with cisplatin administration [Newman *et al.*, 1999]. Although there is a four-fold higher exposure of total-Pt in tumors after SPI-77 compared with cisplatin, this has not translated into antitumor response in clinical trials [Kim *et al.*, 2001], probably due to the lack of release of active unbound cisplatin from the liposome into the tumor extracellular fluid or simply the fact that clinical studies delt with other cancer types.

The anticancer drug, adriamycin (ADR), was incorporated by physical entrapment into polymeric micelles for selective delivery to a murine solid tumor colon adenocarcinoma 26 (C26). In vivo antitumor activity of adriamycin was greatly enhanced by this incorporation into polymeric micelles [Yokoyama *et al.*, 1999]. Using one polymeric micelle delivery system, the tumor completely disappeared at two doses, while free adriamycin exhibited a fair inhibition effect on tumor growth only at the maximum tolerated dose. Biodistribution analysis revealed that the physically entrapped micellar adriamycin accumulated at tumor sites in a highly selective manner. These results indicate that these polymeric micelles are a promising system for delivering hydrophobic anticancer drugs selectively to solid tumor sites using a passive targeting mechanism.

Polymeric micelles incorporating cisplatin (CDDP) were prepared through the polymer-metal complex formation between CDDP and poly(ethylene glycol)-poly(glutamic acid) block copolymers, and their utility as a tumor-targeted drug delivery system was investigated [Nishiyama *et al.*, 2003]. Reduced accumulation of the micelles in normal organs provided high selectivity to the tumor. In vivo antitumor activity assay demonstrated that both free CDDP and the CDDP/m had significant antitumor activity in C26-bearing mice compared with nontreatment, but complete tumor regression was observed only for the treatment with micelles.

The use of proteins or peptides for active liposomal targeting to tumors includes the peptide sequence RGD capable of specific recognition of the $\alpha v\beta 3$-integrin receptor expressed in the neovasculature during angiogenesis of tumor. Thus, the encapsulation of doxorubicin in RGD-addressed liposomes has exhibited superior anticancer efficacy on the C26 colon cancer xenograft model than RGD-nonaddressed liposomes [Schiffelers *et al.*, 2003]. Significant tumor accumulation was also observed using the CC52 antibody directed against rat colon adenocarcinoma as targeting moiety of liposomes [Koning *et al.*, 2003].

The transferrin bearing liposomes also showed the capacity of specific receptor binding and receptor-mediated endocytosis with target colon tumor cells 26 implanted in mice [Ishida *et al.*, 2001].

10. Brain Cancer

Advances in the biology of the blood-brain barrier (BBB) are improving the ability of researchers to target therapeutic peptides, small molecules and other drugs to brain tumors. The BBB is a very specialized system of endothelial cells that separates the blood from the underlying brain cells, providing protection to brain cells and preserving brain homeostasis. In contrast to the open endothelium of the peripheral circulation, the tightly fused junctions of the cerebral capillary endothelium, the anatomic basis for the BBB, essentially form a continuous lipid layer that effectively restricts free diffusional movement of molecules into and out of the brain. Only small, electrically neutral, lipid-soluble molecules can penetrate the BBB by passive diffusion and most chemotherapeutic agents do not fall into this category. Therefore, delivery of drugs to the brain needs a special strategy to bypass the BBB and thus to achieve high intratumoricidal drug concentrations within the central nervous system.

The ability to deliver effective concentrations of therapeutic agents selectively to tumors is however a key factor for the efficacy of cancer therapy. Various strategies have been explored for manipulating the BBB, among them the enhanced drug transport across the barrier by nanoparticulate systems. The selective delivery of nanoparticles to tumor is sometimes achieved due to the "leaky" tumor vasculature, which is known as the EPR effect [Maeda et al., 1989]. The BBB may be partially disrupted and altered by the brain cancer and thus allow the nanoparticles to penetrate into the brain [Neuwelt, 2004].

Others have been found to successfully cross the BBB. The exact mechanism of nanoparticle transport into the brain is not fully understood, but most likely relies on receptor-mediated endocytosis, phagocytosis and/or passive leakage of nanoparticles across defects in the blood–brain barrier [Begley, 2004]. For example, polysorbate-coated nanoparticles are thought to mimic low-density lipoproteins, allowing them to be transported into the brain by the same endocytotic process as low-density lipoproteins undergo at the BBB [Kreuter *et al.*, 2003]. Nanoparticles conjugated with synthetic peptides may be transported across the BBB presumably by a mechanism similar to that of the opioid peptides [Constantino *et al.*, 2005]. The opioid peptides bind to specific receptors on the capillary walls, which help carry the nanoparticles into the brain.

Two different nanoparticle-based therapeutic modalities have been investigated for brain cancer: Chemotherapy and photodynamic therapy.

Chemotherapy has shown a poor outcome due to the low permeability of most anti-cancer agents through the blood–brain barrier. The nanoparticle delivery system has emerged as a promising tool for chemotherapy of brain cancer due to the nanoparticle advantages and the evidence for their ability to cross the BBB.

The therapeutic efficacy of doxorubicin-loaded nanoparticles coated with polysorbate for treating brain cancer was studied in an experimental system based on intracranially implanted 101/8 glioblastoma in rat brains [Steiniger *et al.*, 2004]. The rats treated with doxorubicin nanoparticles showed the most significant increase in survival times compared to the controls.

Besides, muti-functional nanoparticle concept provides another approach for cancer diagnosis and treatment, which integrates the efforts for detection, treatment and follow-up monitoring of the tumor response, leading to decisions about the need for further treatment.

Nanoparticle-based photodynamic therapy for brain cancer has been investigated using polyacrylamide (PAA) nanoparticles [Reddy *et al.*, 2006]. Specifically, a targeted multi-functional nanoplatform combining PDT and MRI with optional hydrophilic coating has been designed for synergistic cancer detection, diagnosis and treatment. The photodynamic therapy agent, Photofrin® is the photosensitizer that is currently approved for clinical use in the USA. The iron oxide was selected as MRI component as the iron oxide-encapsulated PAA nanoparticles had already shown good in vitro and in vivo MRI efficacy. The RGD peptide specifically binds to the αVβ3 integrin that is overexpressed in the tumor vasculature [Arap *et al.*, 1998]. The F3 peptide is a 31-amino acid fragment of human high mobility group protein 2 (HMGN2), which targets to and gets internalized into tumor endothelial cells and certain cancer cells through the nucleolin receptor [Ruoslahti *et al.*, 2005]. Rats bearing intracerebral 9L gliosarcoma tumors survival evaluation resulted in good correlation with diffusion MRI results. There was a statistically significant difference in mean survival time between the non-targeted versus F3-targeted group. However, there was no significant difference in animal survival between the control versus laser-only groups nor between the Photofrin® versus the non-targeted nanoparticle groups.

Transferrin is also a very useful ligand for liposome targeting to tumors. The main advantage of the transferrin receptor as a target arises from its ability to be cell internalized with its specific ligand [Hatakeyama *et al.*, 2004].

Local infiltration of high-grade astrocytomas prevents the complete resection of all malignant cells. It is, therefore, critical to develop delivery systems for chemotherapeutic agents that ablate individual cancer cells without causing diffuse damage to surrounding brain tissue. Sterically stable human interleukin-13 (IL-13)-conjugated liposomes bind efficiently to the brain cancer cells that overexpress the IL-13 receptor alpha2 protein [Madhankumar *et al.*, 2006]. The conjugated liposomes bind to glioblastoma multiforme tissue specimens but not to normal cortex. The therapeutic potential and targeting efficacy of the IL-13-conjugated liposomes carrying doxorubicin was tested in vivo using a s.c. glioma tumor mouse model. Results strongly suggested that IL-13-conjugated liposomes carrying cytotoxic agents are a feasible approach for creating a nanovesicle drug delivery system for brain tumor therapy.

11. Hematological Malignancies

Hematological malignancies are subdivided into two groups: leukemia and lymphoma, which are again subdivided in several types depending on various disease parameters. While leukemia are seizing cells in the blood circulation, lymphoma are found in the lymphatic system. Among lymphomas nanocarrier related drug targeting was only studied to date on non-Hodgkins type.

In the case of acute promyelocytic leukemia, liposomes trapping all-*trans*-retinoic acid have already been studied in clinics. As a derivate of the vitamin A, the drug is hydrophobic and requires a sophisticated formulation for intravenous administration. As the drug induces cytochrome P450 expression but its metabolism occurs by the same enzyme, a long-term therapy leads to observations of lower efficacy. In order to avoid this phenomenon, multilamellar liposomes were designed encapsulating all-*trans*-retinoic acid and commercialized under the name Atragen®. Pharmacokinetic in healthy humans showed a constant drug plasma level for around 15 days during the treatment phase while oral administration led to a reduction of the area under the curve after 9 days [Ozpolat *et al.*, 2003]. Following the administration of Atragen® dermal exfoliation a new adverse effect was observed with around one third of the patients. The analysis of therapeutic efficiency in the treatment of promyelocytic leukaemia patients is rather difficult to estimate due to the varying disease stage in the study. However, no correlation between drug plasma concentration and the number of remissions was found [Estey *et al.*, 1996]. Results indicate the necessity of a more profound study of intratumoral drug concentrations in order to exclude a potential multidrug resistence. In the antitumor therapy of promyelocytic leukaemia all-*trans*-retinoic acid is often associated with other drugs where the potential remains to be analyzed for Atragen®.

Polymeric micelles have been proposed recently as an alternative nanocarrier for the delivery of all-*trans*-retinoic [Kawakami *et al.*, 2005].

Generally, the same mechanisms were observed, but the area under the curve was higher than with liposomes and subsequently the hepatic clearance lowered.

Liposomes (disteroylphosphatidylcholine:cholesterol 2:1) have been developed are carrier system for daunorubicin (DaunoXome®) in the therapy of acute leukemia [Ermacora *et al.*, 2000]. Similar to studies on other cancer types, liposomes diminished adverse effects by their ability to redirect drug distribution within the body. In this study no hepatic toxicity was observed at a dose of 60 mg/m^2 on 3 days, while a distinct dose increase to 150 mg/m^2/d for 3 days was well tolerated however causing several cardiovascular adverse effects during a second treatment cycle [Fassas et al., 2002].

Topoisomerase I inhibitors have shown a certain efficiency in therapy of leukemia. In order to increase the residence time of the drug in the plasma liposomes were prepared from lutotecan. Results in mice showed a reduced clearance from blood permitting a higher drug concentration around the circulating cells [Tomkinson *et al.*, 2003]. However, a phase I study on patients with advanced leukemia did not confirm a correlation between the prolonged residence time in the blood and a potential clinical response or toxicity [Giles *et al.*, 2004].

Anthracycline loaded liposomes (DaunoXome®) were also studied for their clinical effect in the therapy of non-Hodgkin's lymphoma. At a dose of 100 mg/m^2 every third week, 39% of patients suffering from relapsed or refractory lymphoma were responder to the treatment with a mean duration of 19 months [Tulpule *et al.*, 2001]. Under these conditions, 79% of the patients showed adverse effects in form of neutropenia.

A phase II study of low-grade non-Hodgkin's lymphomas with pegylated liposomal doxorubicin led to a stabilisation of the disease in around 70%, however only one third of the patients showed at least a partial response for mean duration of 11 months [Di Bella *et al.*, 2003]. Although these results on the use of anthracyclines can be considered as promising, it seems that pegylated doxorubicin liposomes do not possess a satisfying efficiency. This observed difference may be due to the activity of the drug or the disease stage.

Vincristin oftenly used in the therapy of lymphomas possess a distinct neurotoxicity leading to dose reduction consequently risking low remission and survival rates. As, moreover, vincristin exhibits a half-life of a few minutes, a lipsomal formulation permitted to increase the blood circulation time and also a 50-fold increase of intratumor drug concentration [Gelmon *et al.*, 1999; Hillery *et al.*, 2000]. At the same the liposomes allow to double the dose. A phase II study on relapsed non-Hodgkin's lymphomas demonstrated a significant therapeutic response (41%) with a mean survival of 5.5 months [Sarris *et al.*, 2000].

In the case of lymphoma, in-vivo studies were reported using different types of drug loaded nanocarriers. Their efficiency was based on the prolonged residence time for the proposed etoposide loaded solid lipid nanoparticles [Reddy *et al.*, 2005; Reddy *et al.*, 2006] or hybrid liposomes [Nagami *et al.*, 2006]. The results from animal models are promising although surely requiring clinical results in order to confirm the validity of the therapeutic strategy.

In order to further increase the specificity of the drug delivery in lymphoma therapy monoclonal antibody anti CD19 (determinant of malignant B cells) or Fab' fragments were conjugated to stealth liposomes [Sapra *et al.*, 2004]. The presence of antibodies or Fab fragments reduced the circulation time probably due to the immunogenicity of the surface decorating proteins. The comparison of two different anticancer drugs, vincristin and doxorubicin led the authors to the conclusion that a clinical evaluation of a targeted liposomal vincristin formulation would be most promising.

Another strategy proposed recently is the induction of antitumoral immunity. A subcutaneously injected vaccine is composed of liposomes tumor immunoglobulin protein of the patient and interleukin-2. Interleukin-2 (IL-2) served as adjuvant in order to induce a T-cell response. Liposomes probably provide a depot effect and cause a sustained release of antigen and IL-2 and reach local lymphoid organs after subcutaneous admnistration. Specific anti tumor immunoglobulin protein antibody were detected in 40% of the patients [Neelapu *et al.*, 2004], however the number of patients appeared to be insfficient in order to really correlate immune response and clinical outcome.

Liposomes are the most advanced form in clinical studies of nanocarriers which principally increase the drug concentration in the plasma. Unfortunately, this is not by all means related to a higher antitumor activity. This can be due to several reasons such as multidrug resistence effects on certain drug or difficulties to release the drug from the liposomes into the target tissue. However, the unambiguous benefit from these targeted formulations is the significant reduction of the toxicity (cardiotoxicity, neurotoxicity, myelosuppression). In order to optimize the real therapeutic benefit from the targeted nanocarriers it remains to estimate the intratumoral drug concentration and local availability of the drug at the site of action.

The difficulty in the therapy of the two disease forms is the individually adapted therapeutic doses and treatment scheme. Moreover, in non-Hodgkins lymphoma the potential presence of solid tumors turns things more complicated.

12. Conclusion

In summary, drug loaded nanocarriers should be considered to be very promising. However, it does not becomes clear for an overall comment on whether they efficient or not in cancer therapy. In some cases, a distinct therapeutic progress is visible in others, there is only a palliative effect. In nearly all studies, the presented nanocarriers permit to reduce adverse effects, mainly cardiotoxicity. Surprisingly, no hepatotoxicity was mentioned although long-circulating systems are susceptible to accumulate in the liver.

References

Akamo, Y., Mizuno, I., Yotsuyanagi, T., Ichino, T., Tanimoto, N., Yamamoto, T., Nagata, M., Takeyama, H., Shinagawa, N., Yura, J. and et al. (1994). Chemotherapy targeting regional lymph nodes by gastric submucosal injection of liposomal adriamycin in patients with gastric carcinoma. *Jpn. J. Cancer Res.*, 85, pp. 652-658.

Ameller, T., Marsaud, V., Legrand, P., Gref, R. and Renoir, J. M. (2003). In vitro and in vivo biologic evaluation of long-circulating biodegradable drug carriers loaded with the pure antiestrogen RU 58668. *Int. J. Cancer*, 106, pp. 446-454.

Arap, W., Pasqualini, R. and Ruoslahti, E. (1998). Cancer treatment by targeted drug delivery to tumor vasculature in a mouse model. *Science*, 279, pp. 377–380.

Barraud, L., Merle, P., Soma, E., Lefrançois, L., Guerret, S., Chevallier, M., Dubernet, C., Couvreur, P., Trepo, C. and Vitvitski, L. (2005). Increase of doxorubicin sensitivity by doxorubicin-loading into nanoparticles for hepatocellular carcinoma cells in vitro and in vivo. *J. Hepatol.*, 42, pp. 736-743.

Batist G., Ramakrishman, G., Rao, C. S., Chandrasekharan, A., Gutheil, J., Guthrie, T., Shah, P., Khojasteh, A., Nair, M. K., Hoelzer, K., Tkaczuk, K., Park, Y. C. and Lee, L. W. (2001). Reduced cardiotoxicity and preserved antitumor efficacy of liposome-encapsulated doxorubicin and cyclophosphamide compared with conventional doxorubicin and cyclophosphamide in a randomized, multicenter trial of metastatic breast cancer. *J. Clin. Oncol.,* 19, pp. 1444–1454.

Begley, D. J. (2004). Delivery of therapeutic agents to the central nervous system: the problems and the possibilities. *Pharmacol. Ther.*, 104, pp. 29–45.

Brigger, I., Dubernet, C. and Couvreur, P. (2002). Nanoparticles in cancer therapy and diagnosis. *Adv. Drug Deliv. Rev.*, 54, pp. 631-651.

Chiannilkulchai, N., Ammoury, N., Caillou, B., Devissaguet, J. P. and Couvreur, P. (1990). Hepatic tissue distribution of doxorubicin-loaded nanoparticles after i.v. administration in reticulosarcoma M 5076 metastasis-bearing mice. *Cancer Chemother. Pharmacol.*, 26, pp. 122-126.

Choudhury, A., Charo, J. and Parapuram, S. K. Hunt, R. C., Hunt, D. M., Seliger, B. and Kiessling, R. (2004). Small interfering RNA (siRNA) inhibits the expression of the Her2/neu gene, upregulates HLA class I and induces apoptosis of Her2/neu positive tumor cell lines. *Int. J. Cancer,* 108, pp. 71–77.

Costantino, L., Gandolfi, F., Tosi, G., Rivasi, F., Vandelli, M. A. and Forni, F. (2005). Peptide-derivatized biodegradable nanoparticles able to cross the blood–brain barrier. *J. Control. Rel.*, 108, pp. 84–96.

Di Bella, N. J., Khan, M. M., Dakhil, S. R., Logie, K. W., Marsland, T. A., Weinstein, R. E., Mirabel, M. Y. and Asmar, L. (2003). Pegylated liposomal doxorubicin as single-agent treatment of low-grade non-Hodgkin's lymphoma: A phase II multicenter study. *Clin. Lymphoma*, 3, pp. 235-240.

Dvorak, H. F., Nagy, J. A., Dvorak, J. T. and Dvorak, A. M. (1988). Identification and characterization of the blood vessels of solid tumors that are leaky to circulating macromolecules. *Am. J. Pathol.*, 133, pp. 95-109.

El-Aneed, A. (2004). An overview of current delivery systems in cancer gene therapy. *J. Control. Rel.*, 94, pp. 1-14.

Ermacora, E., Michieli, M., Pea, F., Visani, G., Bucalossi, A. and Russo, D. (2000). Liposome encapsulated daunorubicin (daunoxome) for acute leukaemia. *Haematologica*, 85, pp. 324-325.

Estey, E., Thall, P. F., Mehta, K., Rosenblum, M., Brewer, T. and Simmons, V. (1996). Alterations in tretinoin pharmacokinetics following administration of liposomal all-trans-Retinoic acid. *Blood*, 87, pp. 3650-3654.

Fassas, A., Buffels, R., Anagnostopoulos, A., Gacos, E., Vadikolia, C., Haloudis, P. and Kaloyannidis, P. (2002). Safety and early efficacy assessment of liposomal daunorubicin (DaunoXome) in adults with refractory or relapsed acute myeloblastic leukaemia: A phase I-II study. *Br. J. Haematol.*, 116, pp. 308-315.

Fujisaka, Y., Horiike, A., Shimizu, T., Yamamoto, N., Yamada, Y. and Tamura, T. (2006). Phase 1 clinical study of pegylated liposomal doxorubicin (JNS002) in Japanese patients with solid tumors. *Jpn. J. Clin. Oncol.*, 36, pp. 768-774.

Gabizon, A. (1992). Selective tumor localization and improved therapeutic index of anthracyclines encapsulated in long-circulating liposomes. *Cancer Res.*, 52, pp. 891–896.

Gabizon, A., Catane, R., Uziely, B., Kaufman, B., Safra, T., Cohen, R., Martin, F., Huang, A. and Barenholz, Y. (1994). Prolonged circulation time and enhanced accumulation in malignant exudates of doxorubicin encapsulated in polyethylene-glycol coated liposomes. *Cancer Res.*, 54, pp. 987-992.

Gabizon, A., Peretz, T., Sulkes, A., Amselem, S., Ben-Yosef, R., Ben-Baruch, N., Catane, R., Biran, S. and Barenholz, Y. (1989). Systemic administration of doxorubicin-containing liposomes in cancer patients: a phase I study. *Eur. J. Cancer Clin. Oncol.*, 25, pp. 1795-1803.

Gao, Z., Fain, H. D. and Rapoport, N. (2004). Ultrasound-enhanced tumor targeting of polymeric micellar drug carriers. *Mol. Pharm.*, 1, pp. 317-330.

Garcion, E., Lamprecht, A., Heurtault, B., Aubert-Pouessel, A., Denizot, B., Menei, P. and Benoît, J. P. (2006). A new generation of anticancer, drug-loaded, colloidal vectors reverses multidrug resistance in glioma and reduces tumour progression in rats. *Mol. Cancer Therap.*, 5, pp. 1710-1722.

Gelderblom, H., Verweij, J., Nooter, K. and Sparreboom, A. (2001). E. L. Cremaphor: The drawbacks and advantages of vehicle selection for drug formulation. *Eur. J. Cancer*, 37, pp. 1590–1598.

Gelmon, N. L., Tolcher, A., Diab, A. R., Bally, M. B., Embree, L., Hudon, N., Dedhar, C., Ayers, D., Eisen, A., Melosky, B., Burge, C., Logan P. and Mayer, L. D. (1999). Phase I study of vincristine. *J. Clin. Oncol.*, 17, pp. 697-705.

Giles, F. J., Tallman, M. S., Garcia-Manero, G., Cortes, J. E., Thomas, D. A., Wierda, W. G., Verstovsek, S., Hamilton, M., Barrett, E., Albitar, M. and Kantarjian, H. M. (2004). Phase I and pharmacokinetic study of a low-clearance, unilamellar liposomal mulation of lurtotecan, a topoisomerase 1 inhibitor, in patients with advanced leukaemia. *Cancer*, 100, pp. 1449-1458.

Gordon, A. N., Fleagle, J. T., Guthrie, D., Parkin, D. E., Gore, M. E., Lacave, A. J. (2001). Recurrent epithelial ovarian carcinoma: a randomized phase III study of pegylated liposomal doxorubicin *versus* topotecan. *J. Clin. Oncol.*, 19, pp. 3312-3322.

Hayes, M. E., Drummond, D. C. and Kirpotin, D. B., Zheng, W. W., Noble, C. O., Park, J. W., Marks, J. D., Benz, C. C. and Hong, K. (2006). Genospheres: self-assembling nucleic acid-lipid nanoparticles suitable for targeted gene delivery. *Gene. Ther.*, 13, pp. 646–651.

Heidenreich, A., Sommer, F., Ohlmann, C. H., Schrader, A. J., Olbert, P., Goecke, J. and Engelmann, U. H. (2004). Prospective randomized Phase II trial of pegylated doxorubicin in the management of symptomatic hormone-refractory prostate carcinoma. *Cancer*, 101, pp. 948-956.

Hillery, A. M. (2000). New liposomal formulation of vincristine for non-Hodgkin's lymphoma. *Pharm. Sci. Technolo. Today,* 3, pp. 79-80.

Hitzmann, C. J., Wattenberg, L. W. and Wiedmann, T. S. (2006). Pharmacokinetics of 5-fluorouracil in the hamster following inhalation delivery of lipid-coated nanoparticles. *J. Pharm. Sci.*, 95, pp. 1196-1211.

Hubert, A., Lyass, O., Pode, D. and Gabizon, A. (2000). Doxil (Caelyx): an exploratory study with pharmacokinetics in patients with hormone-refractory prostate cancer. *Anticancer Drugs*, 11, pp. 123-127.

Ibrahim, N. K., Samuels, B. and Page, R., Doval, D., Patel, K. M., Rao, S. C., Nair, M. K., Bhar, P., Desai, N. and Hortobagyi, G. N. (2005) Multicenter phase II trial of ABI-007, an albumin-bound paclitaxel, in women with metastatic breast cancer. *J. Clin. Oncol.*, 23, pp. 6019–6026.

Ishida, O., Maruyama, K., Tanahashi, H., Iwatsuru, M., Sasaki, K., Eriguchi, M. and Yanagie, H. (2001). Liposomes bearing polyethyleneglycol-coupled transferrin with intracellular targeting property to the solid tumors *in vivo*. *Pharm. Res.*, 18, pp. 1042-1048.

Jain, R. K. (2001). Delivery of molecular medicine to solid tumors: lessons from *in vivo* imaging of gene expression and function. *J. Control. Rel.*, 74, pp. 7-25.

Kawakami, S., Opanasopit, P., Yokoyama, M., Chansri, N., Yamamoto, T., Okano, T., Yamashita, F. and Hashida, M. (2005). Biodistribution characteristics of all-trans-Retinoic acid incorporated in liposomes and polymeric micelles following intravenous administration. *J. Pharm. Sci.*, 94, pp. 2606-2615.

Kim, E. S., Lu, C., Khuri, F. R., Tonda, M., Glisson, B. S., Liu, D., Jung, M., Hong, W. K. and Herbst, R. S. (2001). A phase II study of STEALTH cisplatin (SPI-77) in patients with advanced non-small cell lung cancer. *Lung Cancer*, 34, pp. 427–432.

Kong, G., Braun, R. D. and Dewhirst, M. W. (2000). Hyperthermia enables tumor-specific nanoparticle delivery: effect of particle size. *Cancer Res.*, 60, pp. 4440-4445.

Koning, G. A., Morselt, H. W., Gorter, A., Allen, T. M., Zalipsky, S., Scherphof, G. L. and Kamps, J. A. (2003). Interaction of differently designed immunoliposomes with colon cancer cells and Kupffer cells. An *in vitro* comparison. *Pharm. Res.*, 20, pp. 1249-1257.

Kreuter, J., Ramge, P., Petrov, V., Hamm, S., Gelprina, S. E., Engelhardt, B., Alyautdin, R., von Briesen, H. and Begley, D. J. (2003). Direct evidence that polysorbate-80-coated poly(butyl cyanoacrylate) nanopaticles deliver drugs to the CNS via specific mechanisms requiring prior binding of the drug to the nanoparticles. *Pharm. Res.* 20, pp. 409–416.

Kreuter, J., Tauber, U. and Illi, V. (1979). Distribution and elimination of poly(methyl-2-14C-methacrylate) nanoparticle radioactivity after injection in rats and mice. *J. Pharm. Sci.*, 68, pp. 1443-1447.

Krishna, R., and Mayer, L. D. (2000). Multidrug resistance (MDR) in cancer. Mechanisms, reversal using modulators of MDR and the role of MDR modulators in influencing the pharmacokinetics of anticancer drugs. *Eur. J. Pharm. Sci.*, 11, pp. 265-283.

Lacoeuille, F., Hindre, F., Moal, F., Passirani, C., Couturier, O., Calès, P., Le Jeune J. J., Lamprecht, A. and Benoît J. P. (2007) In vivo evaluation of lipid nanocapsules as a promising colloidal carrier for paclitaxel, *Int. J. Pharm.,* in press.

Lamprecht, A. and Benoit, J. P. (2006). Etoposide nanocarriers suppress glioma cell growth by intracellular drug delivery and simultaneous P-glycoprotein inhibition. *J. Control. Rel.*, 112, pp. 208-213.

Law, T. M., Mencel, P. and Motzer, R. J. (1994). Phase II trial of liposomal encapsulated doxorubicin in patients with advanced renal cell carcinoma. *Invest New Drugs*, 12, pp. 323-325.

Lee, C. C., Gillies, E. R., Fox, M. E., Guillaudeu, S. J., Frechet, J. M., Dy, E. E. and Szoka, F. C. (2006). A single dose of doxorubicin-functionalized bow-tie dendrimer cures mice bearing C-26 colon carcinomas. *Proc. Natl. Acad. Sci.*, 103, pp. 16649-16654.

Leighl, N. B., Burkes, R. L., Dancey, J. E., Lopez, P. G., Higgins, B. P., David Walde, P. L., Rudinskas, L. C., Rahim, Y. H., Rodgers, A., Pond, G. R. and Shepherd, F. A. (2003). A phase I study of pegylated liposomal doxorubicin hydrochloride (Caelyx) in combination with cyclophosphamide and vincristine as second-line treatment of patients with small-cell lung cancer. *Clin. Lung. Cancer*, 5, pp. 107-112.

Lu, H., Li, B., Kang, Y., Jiang, W., Huang, Q., Chen, Q., Li, L. and Xu, C. (2007). Paclitaxel nanoparticle inhibits growth of ovarian cancer xenografts and enhances lymphatic targeting. *Cancer Chemother. Pharmacol.*, 59, pp. 175-181.

Lukyanov, A. N., Gao, Z., Mazzola, L. and Torchilin, V. P. (2002). Polyethylene glycol–diacyllipid micelles demonstrate increased accumulation in subcutaneous tumors in mice, *Pharm. Res.*, 19, pp. 1424–1429.

Lyass, O., Uziely, B., Ben Yosef, R., Tzemach, D., Heshing, N. I., Lotem, M., Brufman, G. and Gabizon, A. (2000). Correlation of toxicity with pharmacokinetics of pegylated liposomal doxorubicin (Doxil) in metastatic breast carcinoma. *Cancer,* 89, pp. 1037-1047.

Madhankumar, A. B., Slagle-Webb, B., Mintz, A., Sheehan, J. M. and Connor, J. R. (2006). Interleukin-13 receptor-targeted nanovesicles are a potential therapy for glioblastoma multiforme. *Mol. Cancer Ther.,* 5, pp. 3162-3169.

Maeda, H. and Matsumura, Y. (2000). Tumoritripic and lymphotropic principles of macromolecular drugs. *Crit. Rev. Ther. Drug Carr. Syst.*, 6, pp. 193–210.

Maeda, H., Wu, J., Sawa, T., Matsumura, Y. and Hori, K. (2000). Tumor vascular permeability and the EPR effect in macromolecular therapeutics: a review. *J. Control. Rel.*, 65, pp. 271-284.

Maruyama, K., Ishida, O., Takizawa, T. and Moribe, K. (1999). Possibility of active targeting to tumor tissues with liposomes. *Adv. Drug Deliv. Rev.*, 40, pp. 89-102.

Moghimi, S. M., Hunter, A. C. nd Murray, J. C. (2001). Long-circulating and target-specific nanoparticles: theory to practice. *Pharmacol. Rev.*, 53, pp. 283-318.

Moreira, J. N., Ishida, T., Gaspar, R., and Allen, T. M. (2002). Use of the post-insertion technique to insert peptide ligands into pre-formed stealth liposomes with retention of binding activity and cytotoxicity. *Pharm. Res.*, 19, pp. 265-269.

Muggia, F. M. (1997). Clinical efficacy and prospects for use of pegylated liposomal doxorubicin in the treatment of ovarian and breast cancers. *Drugs*, 54, Suppl. 4, pp. 22-29.

Nagami, H., Matsumoto, Y. and Ueoka, R. (2006). Chemotherapy with hybrid liposomes for lymphoma without drugs in vivo. *Int. J. Pharm.*, 315, pp. 167-172.

Namiki, Y., Namiki. T, Date, M., Yanagihara, K., Yashiro, M. and Takahashi, H. (2004). Enhanced photodynamic antitumor effect on gastric cancer by a novel photosensitive stealth liposome. *Pharmacol. Res.*, 50(1), pp. 65-76.

Neelapu, S. S., Baskar, S., Gause, B. L., Kobrin, C. B., Watson, T. M., Frye, A. R., Pennington, R., Harvey, L., Jaffe, E. S., Robb, R. J., Popescu, M. C. and Kwak, L. W. (2004). Human autologous tumor-specific T-cell responses induced by liposomal delivery of a lymphoma antigen. *Clin. Cancer Res.*, 10, pp. 8309-8317.

Neuwelt, E. A. (2004). Mechanisms of disease: the blood–brain barrier. *Neurosurgery*, 54, pp. 131–142.

Newman, M. S., Colbern, G. T., Working, P. K., Engbers C. and Amantea, M. A. (1999). Comparative pharmacokinetics, tissue distribution, and therapeutic effectiveness of cisplatin encapsulated in long-circulating, pegylated liposomes (SPI-077) in tumor-bearing mice. *Cancer Chemother. Pharmacol.*, 43, pp. 1–7.

Newton, H. B. (2006). Advances in strategies to improve drug delivery to brain tumors. *Expert Rev. Neurother.*, 6, pp. 1495-1509.

Nikitin, A. Y., Juarez-Perez, M. I., Li, S., Huang, L. and Lee, W. H. (1999). RB-mediated suppression of spontaneous multiple neuroendocrine neoplasia and lung metastases in Rb+/− mice. *Proc. Natl. Acad. Sci.*, 96, pp. 3916-3921.

Nishiyama, N., Okazaki, S., Cabral, H., Miyamoto, M., Kato, Y., Sugiyama, Y., Nishio, K., Matsumura, Y. and Kataoka, K. (2003). Novel cisplatin-incorporated polymeric micelles can eradicate solid tumors in mice. *Cancer Res.*, 63, pp. 8977-83.

Noguchi, Y., Wu, J., Duncan, R., Strohalm, J., Ulbrich, K., Akaike, T. and Maeda, H. (1998). Early phase tumor accumulation of macromolecules: a great difference in clearance rate between tumor and normal tissues. *Jpn. J. Cancer Res.*, 89, pp. 307-314.

Numico, G., Castiglione, F., Granetto, C., Garrone, O., Mariani, G., Costanzo, G. D., Ciura, P. L., Gasco, M., Ostellino, O., Porcile, G. and Merlano, M. (2002). Single-agent pegylated liposomal doxorubicin (Caelix) in chemotherapy pretreated non-small cell lung cancer patients: a pilot trial. *Lung Cancer*, 35, pp. 59-64.

O'Brien, M. E., Wigler, N. Inbar, M., Rosso, R., Grischke, E., Santoro, A., Catane, R., Kieback, D. G., Tomczak, P., Ackland, S. P., Orlandi, F., Mellars, L., Alland, L. and Tendler, C. (2004). Reduced cardiotoxicity and comparable efficacy in a phase III trial of pegylated liposomal doxorubicin HCL (Caelyx^{tm}/Doxil$^{®}$) versus conventional doxorubicin for first-line treatment of metastatic breast cancer. *Ann. Oncol.*, 15, pp. 440–449.

Osborne, C., Wilson, P. and Tripathy, D. (2004). Oncogenes and tumor suppressor genes in breast cancer: potential diagnostic and therapeutic applications. *Oncologist*, 9, pp. 361–377.

Otterson, G. A., Villalona-Calero, M. A., Sharma, S., Kris, M. G., Imondi, A., Gerber, M., White, D. A., Ratain, M. J., Schiller, J. H., Sandler, A., Kraut, M., Mani, S. and Murren, J. R. (2007). Phase I study of inhaled Doxorubicin for patients with metastatic tumors to the lungs. *Clin. Cancer Res.*, 13, pp. 1246-1252.

Ozpolat, B. and Lopez-Berestein, G. (2003). Pharmacokinetics of intravenously administered liposomal all-trans-retinoic acid (atra) and orally administered atra in healthy volunteers. *J. Pharm. Pharmaceut. Sci.*, 6, pp. 292-301.

Ozpolat, B., Mehta, K., Tari, A. M. and Lopez-Berestein, G. (2002). All-trans-Retinoic acid-induced expression and regulation of retinoic acid 4-hydroxylase (CYP26) in human promyelocytic leukaemia. *Am. J. Hematol.*, 70, pp. 39-47.

Papahadjopoulos, D., Allen, T. M., Gabizon, A., Mayhew, E., Matthay, K., Huang, S. K., Lee, K. D., Woodle, M. C., Lasic, D. D., Redemann, C. and Martin, F. J. (1991). Sterically stabilized liposomes: improvements in pharmacokinetics and antitumor therapeutic efficacy. *Proc. Natl. Acad. Sci.*, 88, pp. 11460–11464.

Papahadjopoulos, D., Allen, T. M., Gabizon, A., Mayhew, E., Matthay, K., Huang, S. K., Lee, K., Woodle, M. C., Lasic, D. D., Redemann, C. and Martin, M. F. (1991). Sterically stabilized liposomes: improvements in pharmacokinetics and antitumor therapeutic efficacy. *Proc. Natl. Acad. Sci.*, 88, pp. 11460-11464.

Prabha, S. and Labhasetwar, V. (2004). Nanoparticle-mediated wild-type *p53* gene delivery results in sustained antiproliferative activity in breast cancer cells. *Mol. Pharm.*, 1, pp. 211–219.

Reddy, G. R., Bhojani, M. S., McConville, P., Moody, J., Moffat, B. A., Hall, D. E., Kim, G., Koo, Y. E., Woolliscroft, M. J., Sugai, J. V., Johnson, T. D., Philbert, M. A., Kopelman, R., Rehemtulla, A. and Ross, B. D. (2006). Vascular targeted nanoparticles for imaging and treatment of brain tumors. *Clin. Cancer Res.*, 12, pp. 6677-6686.

Reddy, H. L., Adhikari, J. S., Dwarakanath, B. S., Sharma, R. K. and Murthy, R. R. (2006). Tumoricidal effects of etoposide incorporated into solid lipid nanoparticles after intraperitoneal administration in Dalton's lymphoma tumor bearing mice. *AAPS J.*, 2, pp. E254-262.

Reddy, H. L., Sharma, R. K., Chuttani, K., Mishra, A. K. and Murthy, R. R. (2005). Influence of administration route on tumour uptake and biodistribution of etoposide loaded solid lipid nanoparticles in Dalton's lymphoma tumor bearing mice. *J. Control. Rel.*, 105, pp. 185-198.

Romond, E. H., Perez, E. A. and Bryant, J. *et al.* (2005). Trastuzumab plus adjuvant chemotherapy for operable HER2-positive breast cancer. *N. Engl. J. Med.*, 353, pp. 1673–1684.

Rose, P. G. (2005). Pegylated liposomal doxorubicin: optimizing the dosing schedule in ovarian cancer. *Oncologist*, 10, pp. 205-214.

Ruoslahti, E., Duza, T. and Zhang, L. (2005). Vascular homing peptides with cell-penetrating properties, *Curr. Pharm. Des.*, 11, pp. 3655–3660.

Sapra, P., Moase, E. H., Ma, J. and Allen, T. M. (2004). Improved therapeutic responses in a xenograft model of human B lymphoma (Namalwa) for liposomal vincristine *versus* liposomal doxorubicin via Anti-CD19 IgG2a or Fab' fragments. *Clin. Cancer Res.*, 10, pp. 1100-1111.

Sarris, A. H., Hagemeister, F., Romaguera, J., Rodriguez, M. A., McLaughlin, P., Tsimberidou, A. M., Medeiros, L. J., Samuels, B., Pate, O., Oholendt, M., Kantarjian, H., Burge, C. and Cabanillas, F. (2000). Liposomal vincristine in relapsed non-Hodgkin's lymphomas : Early results of an ongoing phase II trial. *Ann. Oncol.*, 11, pp. 69-72.

Schiffelers, R. M., Koning, G. A., Hagen, T. L. M., Fens, M. H. A. M., Schraa, A. J., Janssen, A. P. C. A., Kok, R. J., Molema, G. and Storm, G. (2003). Anti-tumor efficacy of tumor vasculature-targeted liposomal doxorubicin. *J. Control. Rel.*, 91, pp. 115-122.

Schwab, G., Chavany, C., Duroux, I., Goubin, G., Lebeau, J., Helene, C., Saison-Behmoaras, T. (1994). Antisense oligonucleotides adsorbed to polyalkylcyanoacrylate nanoparticles specifically inhibit mutated Ha-ras-mediated cell proliferation and tumorigenicity in nude mice. *Proc. Natl. Acad. Sci.*, 91, pp. 10460–10464.

Shenoy, D. B. and Amiji, M. M. (2005). Poly(ethylene oxide)-modified poly(epsilon-caprolactone) nanoparticles for targeted delivery of tamoxifen in breast cancer. *Int. J. Pharm.*, 293, pp. 261–270.

Singh, M., Ghose, T., Mezei, M., Belitsky, P. (1991). Inhibition of human renal cancer by monoclonal antibody targeted methotrexate-containing liposomes in an ascites tumor model. *Cancer Lett.*, 56, pp. 97-102.

Skubitz, K. M. (2002). Phase II trial of pegylated-liposomal doxorubicin (Doxil) in renal cell cancer. *Invest. New Drugs*, 20, pp. 101-104.

Slamon, D. J., Leyland-Jones, B., Shak, S., Fuchs, H., Paton, V., Bajamonde, A., Fleming, T., Eiermann, W., Wolter, J., Pegram, M., Baselga, J. and Norton, L. (2001). Use of chemotherapy plus a monoclonal antibody against HER2 for metastatic breast cancer that overexpresses HER2. *N. Engl. J. Med.*, 344, pp. 783–792.

Sparreboom, A., Scripture, C. D., Trieu, V., Williams, P. J., De, T., Yang, A., Beals, B., Figg, W. D., Hawkins, M. and Desai, N. (2005). Comparative preclinical and clinical pharmacokinetics of a cremophor-free, nanoparticle albumin-bound paclitaxel (ABI-007) and paclitaxel formulated in Cremophor (Taxol). *Clin. Cancer Res.*, 11, pp. 4136–4143.

Stathopoulos, G. P., Boulikas, T., Vougiouka, M., Deliconstantinos, G., Rigatos, S., Darli, E., Viliotou, V. and Stathopoulos, J. G. (2005). Pharmacokinetics and adverse reactions of a new liposomal cisplatin (Lipoplatin): phase I study. *Oncol. Rep.*, 13, pp. 589-595.

Steiniger, S. C., Kreuter, J., Khalansky, A. S., Skidan, I. N., Bobruskin, A. I., Smirnova, Z. S., Severin, S. E., Uhl, R., Kock, M., Geiger, K. D. and Gelperina, S. (2004). Chemotherapy of glioblastoma in rats using doxorubicin-loaded nanoparticles, *Int. J. Cancer*, 109, pp. 759–767.

Sudimack, J. J., Adams, D., Rotaru, J., Shukla, S., Yan, J., Sekido, M., Barth, R. F., Tjarks, W. and Lee, R. J. (2002). Folate receptor-mediated liposomal delivery of a lipophilic boron agent to tumor cells *in vitro* for neutron capture therapy. *Pharm. Res.*, 19, pp. 1502-1508.

Suzuki, S., Kawakami, S., Chansri, N., Yamashita, F. and Hashida, M. (2006). Inhibition of pulmonary metastasis in mice by all-trans retinoic acid incorporated in cationic liposomes. *J. Control. Rel.*, 116, pp. 58-63.

Tomkinson, B., Bendele, R., Giles, F. J., Brown, E., Gray, A., Hart, K., LeRay, J. D., Meyer, D., Pelanne, M. and Emerson, D. L. (2003). OSI-211, a novel liposomal topoisomerase I inhibitor, is active in SCID mouse models of human AML and ALL. *Leuk. Res.*, 27, pp. 1039-1050.

Torchilin, V. P., Lukyanov, A. N. and Gao, Z. (2003). Papahadjopoulos-Sternberg B. Immunomicelles: targeted pharmaceutical carriers for poorly soluble drugs. *Proc. Natl. Acad. Sci.*, 100, pp. 6039-6044.

Tulpule, A., Rarick, M. U., Kolitz, J., Bernstein, J., Myers, A., Buchanan, L. A., Espina, B. M., Traynor, A., Letzer, J., Justice, G. R., McDonald, D., Roberts, L., Boswell,

W., Nathwani, B. and Levine, A. M. (2001). Liposomal daunorubicin in the treatment of relapsed or refractory non-Hodgkin's lymphoma. *Ann. Oncol.*, 12, pp. 457-462.

Ueno, N. T., Bartholomeusz, C., Xia, W., Anklesaria, P., Bruckheimer, E. M., Mebel, E., Paul, R., Li, S., Yo, G. H., Huang, L. and Hung, M. C. (2002). Systemic gene therapy in human xenograft tumor models by liposomal delivery of the E1A gene. *Cancer Res.*, 62, pp. 6712-6716.

Verdière, A. C., Dubernet, C., Nemati, F., Soma, E., Appel, M., Ferte, J., Bernard, S., Puisieux, F. and Couvreur, P. (1997). Reversion of multidrug resistance with polyalkylcyanoacrylate nanoparticles: towards a mechanism of action. *Br. J. Cancer*, 76, pp. 198-205.

Yan, D. H., Chang, L. S. and Hung, M. C. (1991). Repressed expression of the HER-2/c-erbB-2 proto-oncogene by the adenovirus *E1a* gene products. *Oncogene*, 6, pp. 343–345.

Yokoyama, M., Okano, T., Sakurai, Y., Fukushima, S., Okamoto, K. and Kataoka, K. (1999). Selective delivery of adriamycin to a solid tumor using a polymeric micelle carrier system. *J. Drug Target.*, 7, pp. 171-186.

Chapter 5

NANOTHERAPEUTICS FOR SKIN DISEASES

Nicolas Atrux-Tallau

Laboratoire de Recherche et Développement de Pharmacie Galénique Industrielle, EA 4169, Faculté de Pharmacie, Université Claude Bernard Lyon-I, Lyon (France)

Françoise Falson

Laboratoire de Recherche et Développement de Pharmacie Galénique Industrielle, EA 4169, Faculté de Pharmacie, Université Claude Bernard Lyon-I, Lyon (France)

Fabrice Pirot

Laboratoire de Recherche et Développement de Pharmacie Galénique Industrielle, EA 4169, Faculté de Pharmacie, Université Claude Bernard Lyon-I, Lyon (France)
and
Hôpital Debrousse, Lyon (France)

1. Introduction

In the present chapter, the attention is focused on the details of novel modalities for the treatment of skin diseases by local therapeutics including submicronic drug carriers. Skin drug delivery involving nanocarriers was detailed in first part of the present chapter whereas the treatments of various skin diseases by nanotherapeutics are emphasized in the last section.

The internal structures of the mammalian body are preserved from exogenous aggressions by the skin. The skin is the largest and heaviest organ of the body with a surface area of 1.8-2.0 m^2 and a weight of 5 kg (with blood) [Agache, 2004]. The average thickness of the skin is 1.2 mm. The skin is organized as a stratified tissue composed of three layers, the hypodermis or subcutis (0.1 cm to several cm), the dermis (1.1 mm) and the epidermis (0.1 mm) [Agache, 2004]. Skin appendages (nails, pilosebaceous follicles, eccrine and apocrine sweat glands) complete the structure of this heterogeneous organ. The stratum corneum is the uppermost layer of the epidermis. The adaptation to terrestrial life involves the presence of stratum corneum as a highly impermeable epidermal membrane which restrains both insensitive water loss from the body and the entrance of exogenous substances into the body [Pirot and Falson, 2004].

The stratum corneum exhibits an heterogeneous structure of cornified cells so-called corneocytes, fully keratinized (80% of total protein of stratum corneum) and stacked together to form a 6–10-μm layer in most regions of the body (except in the palms of the hands and soles of the feet) [Kalia *et al.*, 2001; Pirot *et al.*, 1998]. The intercellular lipids bound to corneocytes are arranged as lamellar bilayers [Hill and Wertz, 2003]) composed of cholesterol esters (15%), cholesterol (32%), saturated long-chain fatty acids (16%) and ceramides (37%) [Norlen *et al.*, 1999].

2. Challenging Drug Delivery Across the Stratum Corneum

The treatments of skin diseases by topical application of drug products have several advantages among those e.g., minimal systemic effects, the absence of first-pass effect (or first-pass metabolism) by the liver, the potential drug targeting of skin areas and cutaneous layers. However, the efficiency of topical treatment is related to the penetration and permeation of drugs across the nonviable uppermost layer of the epidermis, the stratum corneum which limits the entry and the subsequent diffusion of exogenous compounds in the viable epidermis and dermis. This resistance of stratum corneum, acting as a membrane

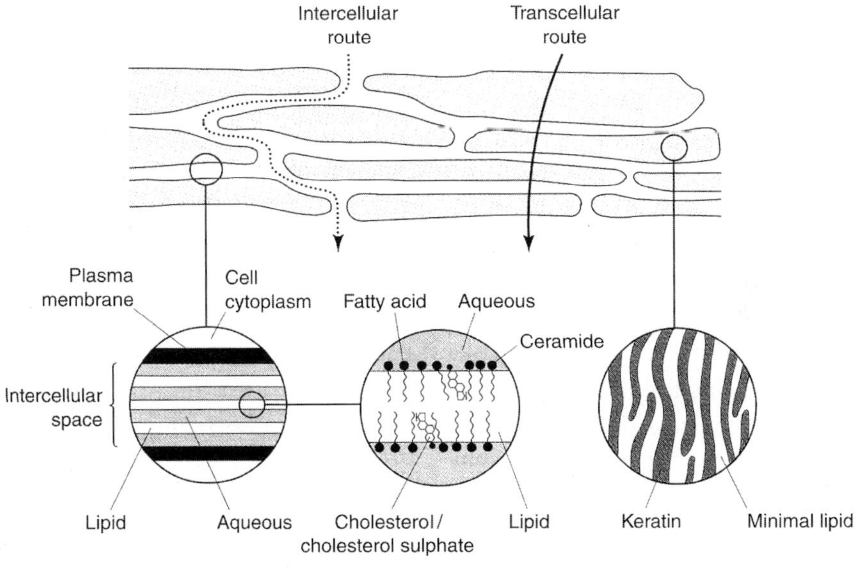

Fig. 1 Schematic representation of the "brick and mortar" model of the stratum corneum and a very simplified lamellar organization of the intercellular domain in which only major stratum corneum lipids are shown including also possible pathways of drug permeation through intact stratum corneum [Moghimi *et al.*, 1996].

against further drug absorption into skin layers, thereby determines the physicochemical properties (e.g., molecular size, solvatochromic parameters, lipophily and related parameters) of drug candidates and vehicles used for topical formulations [Beetge *et al.*, 2000; Karzel and Liedtke, 1989]. Several mathematical models have been proposed to predict percutaneous penetration based on molecular descriptors [Geinoz *et al.*, 2004; Hadgraft, 2004], contribution of free-volume diffusion through lipid bilayers, probabilistic analyses, artificial neural network modelling, fuzzy modelling and biopartitioning micellar chromatography as reported recently [Degim, 2006]. The ultimate goal of these approaches is firstly to develop a predictive global model for skin permeability and secondly to provide mechanistic insight of skin permeability.

In normal stratum corneum, the main route of penetration is predominantly through the intercellular lipids presenting a variety of "tortuous" lipophilic and hydrophilic domains, so that the pathlength of diffusion for a drug is longer that the "transversal" stratum corneum thickness. In spite of its heterogeneous structure often depicted as "brick and mortar" system (Figure 1), the stratum corneum might be assimilated as a homogeneous membrane by considering the intercellular lipids behave as continuum for drug diffusion between the skin surface and the top of living epidermis. The Fick's first law lets to describe the steady state flux of drug (J_{ss}) through the stratum corneum in terms of the partition of the permeate between the skin surface (x = 0) and the vehicle (K), the diffusion coefficient (D) in the intercellular spaces of diffusional pathlength (h), the applied concentration of the permeate in the vehicle (C_{veh}.) and the concentration of the permeate at the bottom of the stratum corneum ($C_{x=h}$) (Eq. 1).

$$J_{ss} = \frac{dQ}{A.dt} = \frac{KD\left(C_{veh} - C_{x=h}\right)}{h} \tag{1}$$

Eq. 1 Where Q is the amount of drug transported by unit of time (t) and surface (A).

One of these demonstrations of the relevance of Fickian model to describe skin absorption was made in the end of 1990s by considering the water and model compound diffusion through the stratum corneum as a function of thickness removed by successive tape-stripping [Pirot *et al.*, 1998; Pirot *et al.*, 1996; Stinchcomb *et al.*, 1999]. The strictly application of Fick's laws of diffusion confirmed (i) the passive process of molecular transport through the stratum corneum, (ii) the invariability of diffusion at any level of depth in the stratum corneum confirming the "apparent" homogeneous pattern of diffusion.

This demonstration was at the origin of further applications for the comparison of cutaneous bioavailability of topical formulations [Kalia *et al.*, 2001].

This reliable, easy-handle methodology opens a window from classical (i.e., simple recording of transient drug amount (Q) in the stratum corneum at defined time exposure expressed as, e.g., Q ($\mu g.cm^{-2}$) $f(t_n)$) to modern (i.e., global analysis of drug concentration (C) profiles as a function of time exposure and position (x) within the membrane, expressed as, e.g., C ($\mu g.cm^{-3}$) $f(t_n, x)$) dermopharmacokinetics.

The lipophilic nature of the intercellular spaces stressed as the major component of skin barrier function determines therefore the physicochemical properties of agents enabling significant diffusion through the stratum corneum. If the diffusion coefficient (*D*) might be regarded as the inherent property of defined compound crossing the stratum corneum, the partition coefficient (*K*) refers to the tendency of vehicle to increase or decrease the amount of compound available at the stratum corneum surface - vehicle interface for ensuing transport. In others words, the vehicle settles on "the dose" of drug at the stratum corneum surface whereas the stratum corneum imposes "the kinetic" of drug distribution within the skin. The latter should be nuanced by the fact that the vehicle is (etymologically) "moving" and might modify the own permeability of the skin barrier. In this case, the vehicle should be re-qualified as "permeate vehicle". The interplay between vehicle-drug-stratum corneum or "permeate vehicle"-drug-stratum corneum has a crucial importance in the relevance of mathematical models for predicting skin absorption based only on molecular descriptors which unfortunately not only privilege the physicochemical properties of drug towards the stratum corneum but also negligee the interaction between the vehicle, the drug and the stratum corneum. The sole couple drug-stratum corneum must be reconsidered and extended to the notion of trio (permeate) vehicle-drug-stratum corneum.

3. Basic Considerations on Nanocarriers for Skin Drug Delivery

Three anatomical locations might be targeted from topical drug delivery namely the skin itself, the deeper tissues (e.g., muscles) for regional delivery, and the systemic circulation (transdermal delivery) [Surber and Smith, 2005]. The present paragraph restricts on the strategies for skin drug delivery in treatment of skin diseases. The achievement of skin drug delivery needs to conciliate two paradoxical terms: firstly, the major barrier of permeation formed by the stratum corneum needs to be circumvented for skin drug delivery (i.e., skin absorption); secondly, the drug deposition within the skin should be ideally accomplished with a restricted percutaneous absorption. At strictly speaking, the terms of this paradox are not solvable, since Fick's laws stipulate that the rate of drug transport is not separable from the gradient of drug concentration. Consequently, if minimal retention in stratum corneum is assumed, then, increasing the skin absorption would imply to favor the percutaneous absorption. Therefore, another strategy for topical drug formulation is required not only to enable the drug transport through the stratum corneum and/or via the follicular pilosebaceous-units for the achievement of skin absorption, but also to limit the extent of percutaneous absorption [Lboutounne *et al.*, 2004a].

In this field, drug carriers as vehicle have been reported in the recent years as one of the most promising strategy to address skin drug delivery. Indeed, the passage of drug loaded particles through the stratum corneum and/or via the follicular ducts might (i) target the drug deposition in specific skin sites, (ii) control and sustain the cutaneous drug release, (iii) protect the drugs against substantial epidermal metabolism, and (iv) reduce the percutaneous absorption [Lboutounne *et al.*, 2004a].

In the next section, the different particulate carriers investigated for skin drug delivery are classified by considering the nature of the supra-molecular structure (e.g., lipid, surfactant, polymer) and the physicochemical characteristics of the colloidal systems (e.g., matrix or shell structure, lamellarity, amorphous or crystalline arrangement, flexibility, deformability, rigidity).

4. Lipid-Based Nanocarriers

4.1. Liposomes and Proliposomes

Liposomes form a class of lipid vesicles which are still considered as a controversial class of dermal and transdermal carriers. Indeed, conventional liposomes only enhanced skin deposition in the uppermost layers of the stratum corneum, thereby restricting percutaneous permeation or systemic absorption of drugs. The lamellarity, size, charge and cholesterol content may also influence the effectiveness of liposomes as skin drug delivery systems. Interestingly, dermal delivery with skin-lipid liposomes was shown to be more effective than delivery with

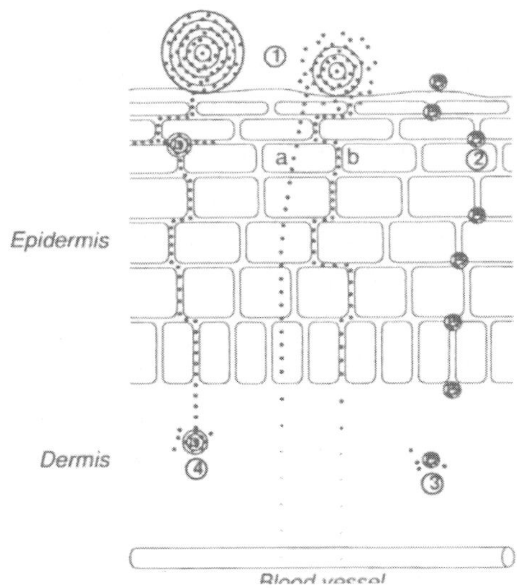

Fig. 2 Hypothetical processes involved in the penetration of liposome-encapsulated drug into the skin [Foldvari *et al.*, 1990].

phospholipid vesicles [Fresno Contreras *et al.*, 2005; Fresta and Puglisi, 1997].

Several mechanisms were suggested for explaining enhanced skin delivery of drugs; among those one distinguishes intact vesicular skin penetration, penetration enhancing effect, the adsorption effect and the penetration of liposomes through the transappendageal route [Elsayed *et al.* 2007]. The detection of unilamellar liposomes in dermis from topical application of multilamellar liposomes was explained conceptually by the lose of external bilayers during penetration (Figure 2). However, unilamellar liposomes exhibited higher promoting skin absorption effect than multilamellar vesicles confirming the dependence of lamellarity and size on skin deposition [Fresta and Puglisi, 1996].

The impact of particle size of liposomes on dermal delivery of drugs was clearly evidenced by [Verma *et al.*, 2003] using confocal laser scanning microscopy enabling the visualization of maximum fluorescence in skin treated by smallest fluorescent lipid vesicles.

4.2. *Transfersomes*

As compared to liposomes defined in the previous section, transfersomes are characterized by the use of edge activator (e.g., sodium cholate, sodium deoxycholate, Tween 80 and Span 80) increasing the elasticity of lipid bilayers [Hiruta *et al.*, 2006]. Therefore, the ultra-deformable vesicles penetrated intact skin along with transdermal osmotic gradients and hydration force [Cevc, 2004; Cevc and Gebauer, 2003]. Cevc *et al.* reported several investigations showing the potential of transfersomes for skin drug delivery and percutaneous absorption [Cevc, 2004; Cevc and Blume, 1992; Cevc and Blume, 2004; Cevc *et al.*, 1997; Cevc *et al.*, 1996; Cevc and Gebauer, 2003; Cevc *et al.*, 1995; Cevc *et al.*, 2002].

4.3. Ethosomes

Ethosomes are lipid vesicles containing high content of ethanol (20-50%) [Godin and Touitou, 2003] acting as drug penetration enhancer and fluidizer for membrane (Figures 3 and 4).

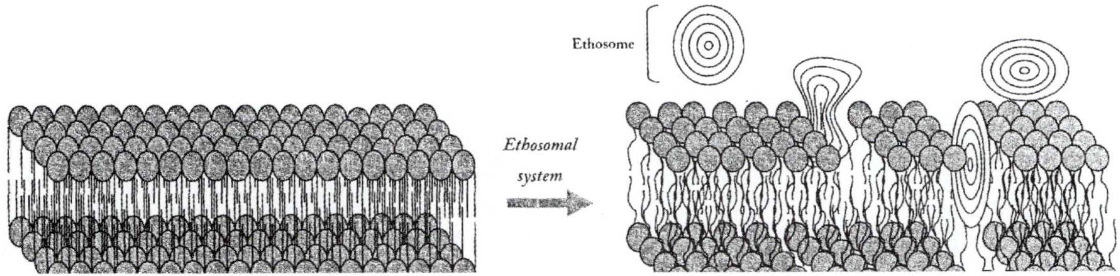

Fig. 3 Proposed mechanism for permeation of molecules from ethosomal system through stratum corneum (SC) lipids: (A) Organized SC lipid bilayers; (B) SC lipid bilayers disturbed by ethanol and penetrated by soft malleable ethosomes [Godin and Touitou, 2003].

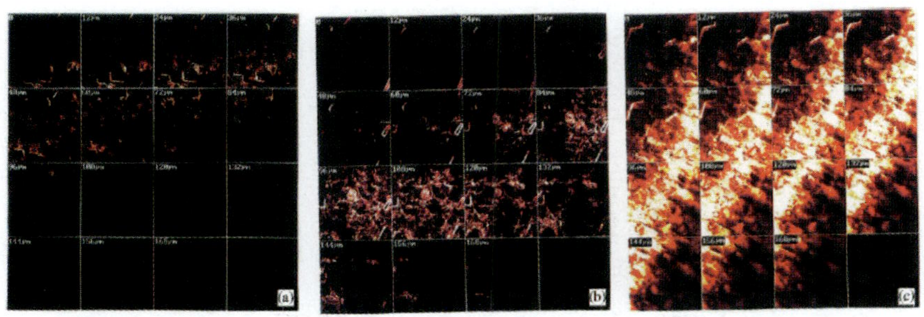

Fig. 4 CLSM micrographs of mouse skin, after application of the fluorescent probe D-289 from: (a) THP liposomes, (b) THP hydroethanolic solution, (c) THP ethosomes from [Dayan and Touitou, 2000].

Due to their high malleability, ethosomes might carrier wide variety of drugs into and through the skin (e.g., trihexyphenidyl [Dayan and Touitou, 2000]; cannabidiol [Lodzki *et al.*, 2003]; ammonium glycyrrhizinate [Paolino *et al.*, 2005]), into fibroblasts (e.g., bacitracin [Godin and Touitou, 2004; Touitou *et al.*, 2001]) as well as into systemic blood circulation (e.g., melatonin [Dubey *et al.* in press]; testosterone [Touitou *et al.*, 2000]). Up-to-date exhaustive review by [Elsayed *et al.* in press] reported *in vivo* studies investigating efficiency and applications of ethosomes as carriers for skin drug delivery, and *in vitro* permeation/deposition studies.

4.4. Aspasomes

Amphiphiles molecules having antioxidant activity (e.g., ascorbyl palmitate) might form multilayered vesicles in combination with cholesterol and charged lipids encapsulating drugs (e.g., azidothymidine) [Gopinath *et al.*, 2004].

Azidothymidine encapsulated in aspasomes (i.e., vesicles made with a mixture of ascorbyl palmitate, cholesterol and dicetyl phosphate) permeated more through excised rat skin than from azidothymidine solution or azidothymidine-ascorbyl palmitate dispersion (Fig. 5).

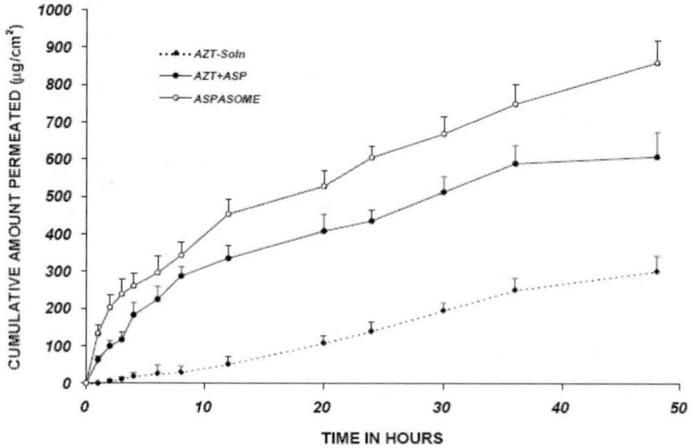

Fig. 5 *In vitro* permeation profiles of azidothymidine (AZT) across excised rat skin following AZT-solution, AZT-ASP dispersion and aspasomal AZT treatments. Adapated from [Gopinath *et al.*, 2004].

4.5. Solid Lipid Nanoparticles

Solid lipid nanoparticles (SLN) constitute a recent alternative to existing derivate-lipid carrier system presenting several advantages including (i) the possibility of controlled drug release and drug targeting, (ii) increased drug stability, (iii) high drug payload, (iv) incorporation of lipophilic and hydrophilic drugs, (v) avoidance of organic solvents, (vi) large scale production and sterilization as reviewed by [Mehnert and Mäder, 2001; Müller *et al.*, 2000]. Conventional production of SLN is made by high-pressure homogenization or modified high shear homogenization and ultrasound techniques [Hou *et al.*, 2003]. Chemical (phospholipid and triglyceride stability) and physical (lipid and dispersion modifications) transformations might change the carrier system properties (e.g., load and release capacity) and consequently the *in vivo* fate of SLN (reviewed by [Heurtault *et al.*, 2003]). Furthermore, alternative colloidal systems might be formed during the production or storage such as micelles, liposomes and drug nanocrystals [Mehnert and Mäder, 2001; Müller *et al.*, 2000].Several technics were proposed to investigate the structures of SLN including photon correlation spectroscopy, laser diffractometry, cryo-field emission scanning electron microscopy, Raman spectroscopy, infrared spectroscopy [Saupe *et al.*, 2006] and differential scanning calorimetry [Castelli *et al.*, 2005].

SLN size was correlated with lipid content, smallest sizes being obtained with low lipid content (up to 5%) [Mehnert and Mäder, 2001] which confers to the final formulation a low viscosity unsuitable for topical use. A contrario, increasing lipid content improving viscosity of dermal product results in an increase of SLN size. The topical administration of an oil/water cream enriched with 4% SLN increased skin hydration (by occlusion effect, Figure 6) of 31% after 4 weeks and exhibited significant photoprotective properties [Wissing and Müller, 2003], whereas other studies showed that sunscreen oxybenzone loaded SLN was released and penetrated into human skin more quickly and to a greater extent from conventional emulsions [Wissing and Müller, 2002b]. The occlusion effect of SLN was shown related to the degree of cristallinity of lipid nanoparticles [Wissing and Müller, 2002a]. SLN was found to modulate the skin penetration of drugs [Müller *et al.*, 2002], including topical glucocortocoids [Maia *et al.*, 2000; Maia *et al.*, 2002], retinol, isotretinoin [Liu *et al.*, 2007], podophyllotoxin [Chen *et al.*, 2006], clotrimazole [Souto *et al.*, 2004].

During storage of SLN, lipids tend to form a highly ordered crystal leading to drug expulsion and limited drug loading [Müller *et al.*, 2002]. The crystallisation (or re-crystallisation after sterilisation by heat) of lipids might be circumvented by blending solid lipids with liquid lipids

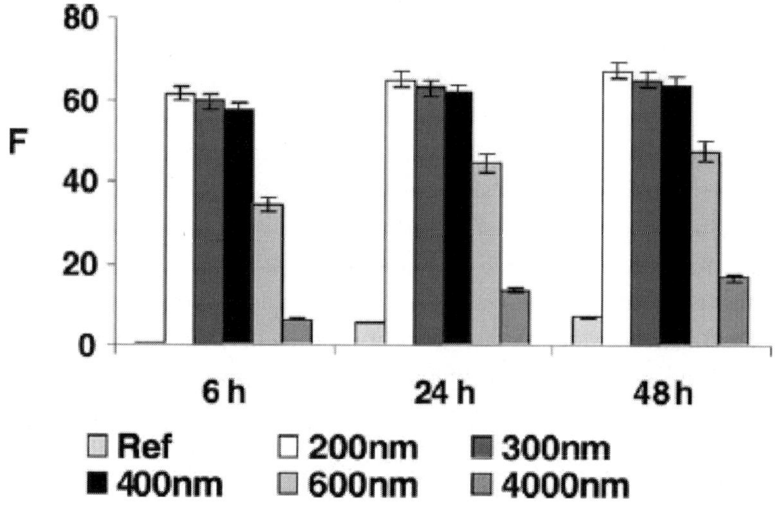

Fig. 6 Occlusion factor F in dependence on the time and as a function of the particle size of the SLN dispersions [Wissing and Müller, 2003].

(e.g., oleic acid, [Hu *et al.*, 2005]) leading to a new concept of SLN so-called nanostructured lipids carriers (NLC).The lipid matrix is solid although characterized by a melting point depression compared to the original solid lipid. Increased indomethacin encapsulation in NLC was shown by differential scanning calorimetry as compared to that determined in SLN [Castelli *et al.*, 2005]. In the same field, higher entrapment efficiency of clotrimazole was shown in NLC exhibiting faster release profile in comparison to SLN with the same lipid concentration and lower occlusive capacity [Souto *et al.*, 2004]. Similar findings were found by [Hu *et al.*, 2006] with monostearin NLC incorporating clobetasol propionate showing that the drug release rates were increased with rising the fraction of liquid lipids (i.e., caprylic/capric triglycerides). In recent publication, the cytoxicity of SLN was reduced by using semi-synthetic glycerides or hard fat instead of stearic acid [Weyenberg *et al.*, 2007].

5. Surfactant-based Nanocarriers

5.1. Niosomes, Proniosomes and Cubosomes

Since 1990's, non-ionic surfactant vesicles (NSV) also referred to as niosomes had been reported as an alternative to liposomes as drug carriers [Manconi *et al.*, 2006]. The advantages of NSV include higher purity and stability of non-ionic surfactants compared to phospholipids (i.e., degradation by hydrolysis or oxidation) [Vora *et al.*, 1998], (ii) the feasibility of large-scale production avoiding the use of organic solvents [Fang *et al.*, 2001b], (iii) the adjustment of drug release rates and drug targeting (e.g., drug deposition in the pilosebaceous unit [Tabbakhian *et al.*, 2006]) by modification of their composition or surface [Alsarra *et al.*, 2005; Schreier and Bouwstra, 1994]. Niosomes are unilamellar or multilamellar vesicles incorporating a wide variety of hydrophilic and hydrophobic drugs attributable to their amphiphile structure serving thus as a solubilizing matrix, as drug deposition increasing residence time of drugs in the stratum corneum and epidermis [Manconi *et al.*, 2006], as penetration enhancers, or as a rate-limiting membrane barrier for the modulation of systemic absorption of drugs via the skin [Schreier and Bouwstra, 1994]. Niosomes are obtained from the hydrated mixtures of cholesterol and non-ionic surfactants such as monoalkyl or dialkyl polyoxyethylene ether, which might be interestingly substituted by less toxic, highly biodegradable sugar-based surfactants such as alkyl polyglucoside surfactants [Manconi *et al.*, 2006; Mura *et al.*, 2007]. Proniosomes, dehydrated form of niosomes, constitute, from a technical point a view, an interesting option for (i) NSV storage [Mura *et al.*, 2007] and (ii) topical treatment under occlusive conditions [Alsarra *et al.*, 2005; Vora *et al.*, 1998]. The topical use of drug loaded niosomes might improve the cutaneous availability and therefore reduce the dose necessary for determining a therapeutic effect and the dose-dependent side-effects like irritation and staining [Agarwal *et al.*, 2001; Alsarra *et al.*, 2005; Fang *et al.*, 2001a; Manconi *et al.*, 2006; Tabbakhian *et al.*, 2006].

Surfactants may form organized phases of molecular aggregates as lamellar, hexagonal and cubic liquid crystals according to their concentration into the formulation [Brinon *et al.*, 1999]. Vehicle-skin interactions of liquid poly(oxyethylene)-dodecyl ether crystals on percutaneous absorption of liposoluble (octyl methoxycinnamate) and hydrosoluble (benzophenone-4) sunscreens was investigated through pig skin and showed that the percutaneous absorption of benzophenone-4 was strongly dependent on the nature of liquid crystals. In the same field, [Esposito *et al.*, 2005] showed that the incorporation of indomethacin in monoglyccride/poloxamer 407/water system presenting 72% of cubosomes and vesicles in the nanometer size (plus 28% of larger irregular particles) improved the index of inhibition of the erythema correlated with higher amount of drug into the stratum in comparison with free indomethacin formulations.

5.2. Microemulsions

Microemulsions are defined as clear and thermodynamically stable isotropic mixture of oil, water and surfactant/co-surfactant [Danielsson and Lindman, 1981], obtained almost spontaneously taking account the zero or very low interfacial tension between dispersed and continuous phases. Additional properties include low viscosity with Newtonian behavior, high surface area, high solubilization capacity and very small droplet size [Kogan and Garti, 2006]. Therefore, those systems should not be confused with submicronic emulsions produced after extrusion of macroemulsions through nanofilters. The use of microemulsions as transdermal drug delivery vehicles was recently and extensively reviewed by [Kogan and Garti, 2006] and [Kreilgaard, 2002]. Skin drug delivery from microemulsions is influenced by multiple factors depending on the characteristics of the applied constituents so that the impact of separate components on the extent of drug penetration and permeation through the skin is difficult to appreciate. However, the characteristic ultra-low interfacial tension encountered in all microemulsions would ensure an excellent surface contact between the skin and the vehicle over the entire application area [Kreilgaard, 2002] facilitation further skin penetration. Furthermore, partition between relative hydrophilic vehicle, i.e. microemulsion and the lipophilic stratum corneum is favored by the internal fluctuations of aqueous phase, lipophilic phase and surfactants embedding lipophilic and hydrophilic drugs [Kreilgaard, 2002]. The formation of super-saturated microemulsions during skin exposure was also reported as favorable condition for increasing cutaneous drug delivery [Kemken *et al.*, 1991a; Kemken *et al.*, 1991b; Kemken *et al.*, 1992].

6. Polymer-based Nanocarriers

6.1. Nanoparticles

The effect of the inclusion of drugs in polymeric nanoparticulate carriers on transdermal drug delivery was reviewed twenty years ago (Kreuter, 1988). Since, several reports emphasized site-specific drug delivery of polymeric micro- and nanoparticles in pilosebaceous structures [Rolland, 1993; Lademann *et al.* 2007; Lboutounne *et al.*, 2004a; Meidan *et al.*, 2005; Vogt *et al.*, 2006] (Figure 7) and dermatoglyphs [Luengo *et al.*, 2006].

Furthermore, the uptake of polymeric nanoparticles by epidermal cells was found dependent on their size [Vogt *et al.*, 2006], whereas benzopsoralen-loaded poly(D,L-lactic-co-glycolic acid) nanoparticles were endocyted by the majority of the cells present in the rat cell exsudate confirming the potential of such carriers to target cellular structures [Gomes *et al.*, 2007].

Therefore, polymeric nanoparticles were suggested to increase the skin drug concentration within pilosebaceous units, to improve the therapeutic index of certain drugs (e.g., adapalene, [Rolland, 1993],

5-fluorouracil, [Simeonova *et al.*, 2003], all-*trans* retinoic acid, [Yamaguchi *et al.*, 2005]), to avoid degradation of drugs at the skin surface and to control the drug release onto the stratum corneum and into the hair follicles [Lboutounne *et al.*, 2004a].

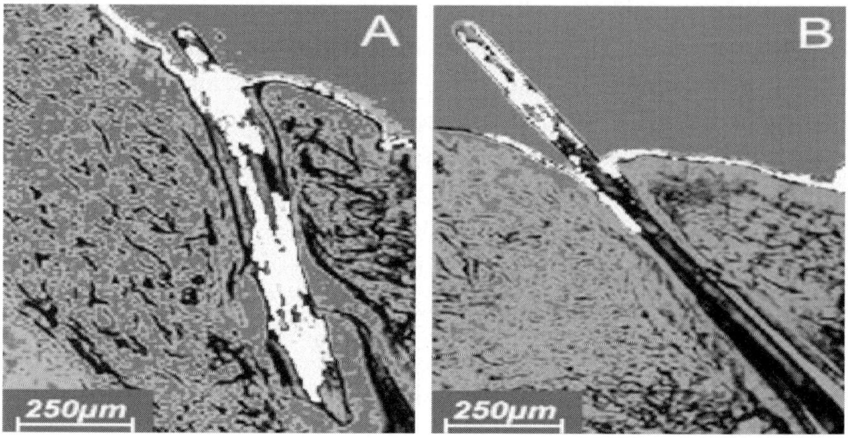

Fig. 7 Superposition of a transmission and fluorescent image, demonstrating the in vitro penetration of the dye-containing formulation into the hair follicles of porcine skin after application of a massage. (A) Dye in particle form. (B) Dye in non-particle form [Lademann *et al.* 2007].

7. Dermatological Treatments

7.1. Alopecia

Alopecia is commonly divided in three types: (i) androgenetic alopecia, a common form of hair loss in both men and women, in a well-defined pattern, beginning above both temples, (ii) alopecia areata, which involves the loss of some of the hair from the head, it was thought to be an autoimmune disease in which the body mistakenly treats its hair follicles as foreign tissue and suppresses or stops hair growth in spotted area; and (iii) alopecia totalis, which involves the loss of all head hair, to the most extreme form, alopecia universalis, which involves the loss of all hair from the head and the body. Depending alopecia type, treatments will be very different and in some case no treatment will be available.

Minoxidil is a vasodilator and originally was exclusively used as an oral drug (Lonoten®) to treat high blood pressure, modulating potassium channel and may be a nitric oxyde agonist. It was, however, discovered to have the interesting side effect of hair growth and reversing baldness. Thus, it appeared quickly commercial solution containing up to 5% of minoxidil. However treatments are long (4 to 12 months with a continuous twice-daily application) weak (efficacy in 30% of subjects) and temporary suppressive.

It has been demonstrated that the ethosomal system dramatically enhanced the skin permeation of minoxidil *in vitro* compared with either ethanolic or hydroethanolic solution or phospholipid ethanolic micellar solution of minoxidil [Touitou *et al.*, 2000]. Mura studied skin penetration and permeation of multilamellar liposomes, niosomes and propylene glycol-water-ethanol solution (control) loaded with minoxidil. They pointed out that penetration of minoxidil in epidermal and dermal layers was greater with liposomes than one with niosomal formulations and the control solution. No permeation of minoxidil through the whole skin thickness was detected suggesting low systemic side-effects [Mura *et al.*, 2007].

7.2. Acne

Acne vulgaris is the most common disease of the skin which can be classified as types I–IV, inflammatory versus noninflammatory, comedonal, comedopapular, papular, papulopustular, pustular, and "cystic" or nodular (even nodular-cystic) [Shalita, 2004].

7.2.1. Tretinoin and Isotretinoin

All-*trans*-retinoic acid or tretinoin is commonly used topically at 0.025%-0.1% in creams, gels, solutions or lotions for the treatment of acne vulgaris. Previous reports showed that the encapsulation of tretinoin in liposomes reduced skin irritancy [Patel *et al.*, 2000], improved patient compliance (less burning and erythema), allowed to decrease the concentration of the active agent without a decline in efficacy [Schäfer-Korting *et al.*, 1994]. Efficiency of retinoid liposomal formulations will be associated (i) to the containment of irritant agents in either lipid bilayers or aqueous core [Date *et al.*, 2006], (ii) to the target of sebaceous glands via the follicular route (e.g., 13-*cis*-retinoic acid or isotretinoin, [Tschan *et al.*, 1997]), (iii) to the drug retention by the viable skin [Masini *et al.*, 1993; Fresno Contreras *et al.*, 2005]. Tretinoin delivery in skin was clearly found dependent of the membrane charge [Kitagawa and Kasamaki, 2006], composition and lamellarity of liposomes [Sinico *et al.*, 2005].

Solid lipid nanoparticles were also considered as a possible carrier to limit side-effects of vitamin A derivatives in acne treatments, notably sustain release may avoid skin irritancy. Retinol and retinyl palmitate were loaded, separately, in SLN dispersions in hydrogel or oil/water cream to mimic cosmetical – pharmaceutical forms. It appears that SLN loaded were compatible with these formulations, especially uncharged or weekly charged polymers, like xanthan gum or cellulose, which revealed a low interaction potential and were, therefore, well suited to achieve the desired thickening action [Jenning *et al.*, 2000b]. Release profile of such systems was due to the transition from metastable SLN system to a stable state (crystal order and density increased whereas amorphous regions and crystal defects decreased). Such modification promoted by e.g. surfactants or thickening agent determined either a rapid drug expulsion (i.e., characterised by a burst effect) and a higher bioavailability, or a slow drug expulsion and subsequent sustained release [Jenning *et al.*, 2000b]. Distribution of SLN, located in upper parts of the skin, decreased available drug for further penetration, avoiding systemic side-effects [Jenning *et al.*, 2000a; Jenning *et al.*, 2000b; Liu *et al.*, 2007]. In another study, the delivery of tretinoin loaded liposomes and niosomes through the skin was found higher than from methanolic tretinoin solution. By varying the structure and/or bilayer composition of vesicle dispersions, the skin drug absorption was modulated [Manconi *et al.*, 2002] as well as the photoprotection of tretinoin improved in unilamellar vesicles [Manconi *et al.*, 2003]. Finally, hydrophilic surfactants were found rising percutaneous absorption whereas less hydrophilic surfactants enhanced skin drug deposition [Manconi *et al.*, 2006]. The delivery of retinoic acid studied from microemulsions was found partly dependent on the formation of ion pair within the vehicle exhibiting higher hydrophobicity which increased drug retention into the epidermis (maximum penetration up to 100 nm) and reduced the permeation through the skin. This suggested that microemulsions may optimise skin drug targeting and reduce further systemic absorption [Trotta *et al.*, 2003].

7.2.2. Cyproterone Acetate

Cyproterone acetate (CPA) is a synthetic derivative of 17-hydroxyprogesterone, and acts as an androgen receptor antagonist used for the treatment of hirsutism and acne vulgaris [Iraji *et al.*, 2006]. The use of topical CPA was investigated to minimize side effects linked to oral treatment. In this field, [Gruber *et al.*, 1998] compared the effectiveness of a lotion of 20 mg CPA in liposomes to an oral daily treatment (2 mg CPA and 35 μg ethinylestradiol) in women with acne. After 3 months of treatment, mean facial acne grade and lesion counts were comparable in the topical CPA and oral medication. Interestingly, the serum level of CPA after topical drug delivery was found ten times lower than that found after oral administration, thus reducing the risk of adverse effects and avoiding high serum CPA concentrations [Gruber *et al.*, 1998]. In a recent study, [Stecova *et al.*, 2007] compared CPA penetration into excised human skin treated by CPA 0.05% loaded SLN, NLC, nanoemulsion and micropheres and demonstrated a drug targeting within skin tissue and likely minimal systemic absorption.

7.2.3. Benzoyl Peroxide

Benzoyl peroxide is an effective topical agent at 2.5%-10% in gels, lotions and dermatological soaps in the treatment of acne. As mentioned for retinoid compounds, topical application of benzoyl peroxide is followed by local irritation and burning as major side effects limiting, again, patient compliance. Liposomal gel of benzoyl peroxide exhibited (i) a lower drug release than benzoyl peroxide dispersed in liposomes, (ii) a reduced local irritation relative to its non liposomal benzoyl peroxide gel, (iii) and also an improved clinical efficacy in the treatment of acne [Patel *et al.*, 2001]. Furthermore, phospholipid liposome formulation of benzoyl peroxide showed a significantly greater antibacterial efficacy for Propionibacteria and Micrococcaceae [Fluhr *et al.*, 1999]. Combination of both tretinoin and benzoyl peroxide in liposomal dispersion showed a synergistic effect in treating all types of acne lesions in addition to a reduction in the duration of therapy as compared to tretinoin and benzoyl peroxide alone [Patel *et al.*, 2001].

7.2.4. Antibiotics: Lincosanides

Clindamycin is an antibiotic belonging to lincosanide group used at 1% in a hydro-alcoholic solution or in a gel for acne treatment. [Skalko *et al.*, 1992] reported that clinical treatment of acne vulgaris with a lotion of liposomal clindamycin showed better efficacy than non-liposome lotion forms. These results were confirmed by [Honzak and Sentjurc, 2000] who showed, in a double-blind clinical study, that (multilamelar) liposome-encapsulated 1% clindamycin solution was therapeutically superior over conventional 1% clindamycin solution in the treatment of acne vulgaris.

7.3. Psoriasis

Psoriasis is a chronic auto-immune disease affecting the skin and joints and characterized (i) clinically, by erythemato-squamous (sharply circumscribed salmon pink) patches or plaques covered by silvery scaling and a chronic recurrent course, (ii) histologically, by the hyperproliferation of the epidermis, elongated and prominent blood vessels and a thick perivascular lymphocytic infiltrate [de Rie *et al.*, 2004]. Treatments include topical (tar, sulphur, salicylate, dermocorticoids, calcipotriol, dithranol), systemic (retinoids, e.g., tazaroten, acitretin; methotrexate, cyclosporine A) and PUVA (i.e, psoralens and ultraviolet A) therapies. Topical cares are preferred in patients with limited lesions, whereas systemic treatments should be reserved for extensive psoriasis and for failure after well-conducted local care.

7.3.1. Vitamin D3 Analogues

Three vitamin D3 analogues are available for the topical treatment of psoriasis during the last decade: calcitriol, calcipotriol and tacalcitol [van de Kerkhof, 2001]. Calcitriol (Silkis®, ointment at 3μg/g), calcipotriol (Daivonex® ointment or cream at 50 μg/g) Diavobet® ointment at 50 μg/g plus betamethasone) and tacalcitol (Apsor®, ointment at 4 μg/g) suppress inflammation and hyperproliferation and promote normal epidermal differentiation in psoriatic skin [Korbel *et al.*, 2001]. Vitamin D3 analogues show the same efficacy as potent topical corticosteroids and do not produce skin atrophy during long-term therapy [Fogh and Kragballe, 2004]. Merz and Sternberg [1994] studied the effects of the incorporation of vitamin D3 analogues in liposomes made of dimyristoyl-glycero-phosphocholine and egg phosphatidylcholine, showing that 80% of calcitriol and calcipotriol was included into the lipid bilayers. Recently, [Prufer and Jirikowski, 1996] demonstrated that the entrapment of vitamin D3 analogues in liposomes enhanced the antiparakeratotic effect in mouse tail test as compared to that of currently available commercial preparations. Skin irritation and hypercalcemia as side-effects of vitamin D3 analogue topical treatments might be thus circumvented by use of liposomal formulations.

7.3.2. Dithranol

Dithranol (1,8-dihydroxy-9-anthrone, MW: 226.23 g/mol; Log$P$ = 4.16), first synthesized in 1916 have since been in clinical use in the treatment of psoriasis [Agarwal *et al.*, 2001]. Dithranol is formulated in ointment at 0.35% associated with salicylic acid (0.3%-1%) and tar (0.3%). However, the application of dithranol elicits severe side-effects such as irritation, burn, staining and necrosis on the normal as well as the diseased skin [Agarwal *et al.*, 2001]. Agarwal *et al.* [2002] prepared a novel, aqueous gel-based, liposome-entrapped formulation of dithranol. Preliminary observations showed effective clearance of lesions in five of nine patients treated by liposomal dithranol gel. Furthermore, there were no reports of lesional or perilesional irritation, and only one patient showed faint brown staining of the skin, which was completely and rapidly reversible [Agarwal *et al.*, 2002]. Saraswat *et al.* [2002] reported that the liposomal gel exhibited superior washability than Derobine® ointment which may potentially increase the acceptability of dithranol amongst psoriasis patients.

Coevaporation of dithranol and polyvinylpyrrolidone in various fraction was found effective in formulating homogeneous aqueous colloidal dispersions with an average particule size less than 0.2 μm [Delneuville *et al.*, 1998].

7.3.4. Dermocorticoids

Ultra-high-potency betamethasone dipropionate with propylene glycol and clobetasol propionate are used at 0.05% in cream or in gel. The use of liposomes to increase the delivery of betamethasone dipropionate to the epidermis increased the antiinflammatory action but not the antiproliferative effect suggesting that liposome encapsulation may improve the benefit-risk ratio in eczema. [Korting *et al.*, 1990]. Skin-lipid (unilamellar) liposomes (~100 nm) of hydrocortisone, betamethasone and triamcinolone acetonide prepared with bovine brain ceramides, cholesterol, palmitic acid and cholesteryl sulphate provided (i) higher epidermal and dermal build up, (ii) higher skin blanching effect, (iii) smaller drug levels in the blood and urine than those determined from control formulation ointment and phospholipid-based liposome formulation [Fresta and Puglisi, 1997]. This behaviour was probably due to an almost complete incorporation of skin-lipid liposomes into and/or mixing with the skin lipids [Fresta and Puglisi, 1997].

7.3.5. Psoralens

Three psoralens (5-methoxypsoralen, 8-methoxypsoralen and 4,5',8-trimethylpsoralen) are used in combination with near-ultraviolet (320-400 nm) light for the treatment of vitiligo, psoriasis, cutaneous T-cell lymphoma, alopecia areata, eczema, and other skin diseases in phototherapy [Pathak and Fitzpatrick, 1992; Potapenko and Kyagova, 1998]. 5-methoxypsoralen and 8-methoxypsoralen doses are delivered *per os* as a function of body weight, whilst 8-methoxypsoralen is applied topically from a hydro-alcoholic solution at 0.1%-0.75%. The topical use of this agent is not associated with adverse systemic symptoms such as nausea [Lebwohl *et al.*, 2005]. In order to improve skin delivery of 4,5',8-trimethylpsoralen, [Lboutounne *et al.*, 2004b] compared the skin penetration and permeation of the photosensitising drug from ethanol solution, liposomal and nanoparticles suspensions. Cutaneous delivery was improved from 4,5',8-trimethylpsoralen colloidal suspensions concomitant to a minimal percutaneous absorption.

Microemulsions system of 8-methoxypsoralen enhanced total penetration through the skin by order of 1.9 – 4.5, as compared with isopropyl myristate [Baroli *et al.*, 2000].

7.3.6. Cyclosporine A

Cyclosporin A is a nonpolar cyclic oligopeptide (11 amino acids) exhibiting immunosuppressive properties showing remarkable efficacy in psoriasis (2.5 mg/kg/day as initial dose).

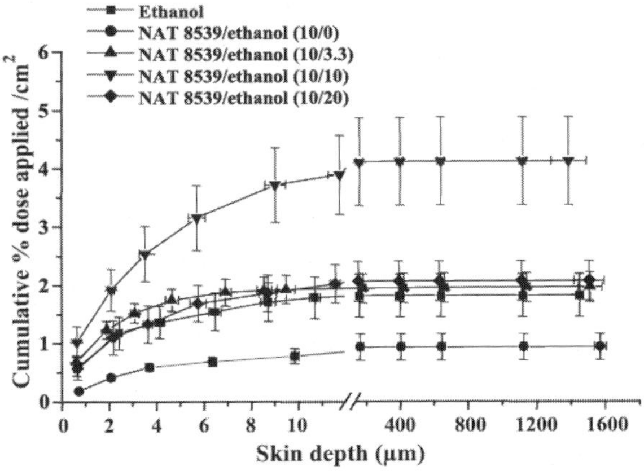

Fig. 8 Skin depth profile of radioactively labeled cyclosporine A from different formulations across human abdominal skin after 6 h of non-occlusive application (expressed as percentage of dose applied/cm^2 ± SE). [Verma and Fahr, 2004].

Systemic administration of cyclosporine A is associated with serious side effects, especially nephrotoxicity. Topical delivery of cyclosporine A is impeded by its physicochemical properties (i.e. high lipophily and molecular mass). [Verma and Fahr, 2004] reported that the dispersion of cyclosporine A in lipid vesicles (with or without ethanol) increased the skin penetration of the drug as compared to that determined with ethanol solution (Figure 8).

7.3.7. Methotrexate

Methotrexate is a folic acid antagonist with antineoplastic activity frequently and effectively used for the treatment of severe, recalcitrant psoriasis in adult by oral (5-7.5 mg as a single-dose /week,) administration over long periods of time. In order to reduce the risk of systemic toxicity (e.g., hepatotoxicity and bone marrow suppression) and also to avoid first-pass degradation or metabolism in the gastrointestinal tract or liver, topical application of methotrexate was investigated. [Wong *et al.*, 2005] reported that methotrexate in liposomal formulation as penetration enhancer (i.e., methotrexate was not encapsulated in liposomes) coupled with electroporation procedure provided a significant transdermal drug delivery within a short application time. *Ex vivo* experiments showed that methotrexate amounts permeated through pig skin were three- to four-fold higher using transfersomes (i.e., deformable liposomes containing soybean lecithin or hydrogenated lecithin and dipotassium glycyrrhizinate as surfactant) compared to those from water solution or normal liposomes [Trotta *et al.*, 2004]. Interestingly, up to 50% of the administered liposomal dose was found in the skin after 24-h exposure.

Methotrexate delivery to the skin from hydrogel by iontophoretic assay or passive diffusion from microemulsions was more effective than passive diffusion from aqueous solutions [Alvarez-Figueroa and Blanco-Mendez, 2001]. Interestingly, in this study, the amounts of methotrexate remaining in the skin were the same from iontophoretic delivery or microemulsions.

7.4. Eczema

Eczema regroups different persistent or recurring skin disorders, characterized by redness, skin oedema, itching and dryness. Etiology of eczema may be allergic or non-allergic related. Topical treatments include skin moisturizing, antihistaminic, dermocorticosteroids with associated side-effects (e.g., skin atrophy, local immunosuppression). As an alternative, immunomodulators (e.g. tacrolimus) have been proposed to suppress local immune response in allergic eczema.

7.4.1. Dermocorticoids

Unilamellar liposomes obtained with human skin-lipids were proposed for the topical administration of hydrocortisone, betamethasone or triamcinolone [Fresta and Puglisi, 1997]. In this study, the therapeutic effectiveness was evaluated by the measurement of the blanching effect following UV-induced erythema. Corticosteroid-loaded liposomes induced a stronger vasoconstriction than in ointment dosage form correlated with higher drug deposition into the epidermis and dermis.

Dermocorticoids have also been loaded in transfersomes and evaluated for their potentiality in skin delivery [Cevc *et al.*, 1997]. The biodistribution and pharmacokinetics of topically or intravenously administred hydrocortisone, dexamethasone, and triamcinolone-acetonide in very deformable vesicles were compared [Cevc *et al.*, 1997]. Authors concluded that transfersomes ameliorated the targetability of all tested compounds into the viable skin or throughout the body [Cevc *et al.*, 1997]. These deformable vesicles allowed a well-controlled topical skin treatment limiting the incidence of side-effects and improving the therapeutic effect of low potent drug (e.g., hydrocortisone). Transfersomes reduced the frequency of application and simplify the dose regimen protocole [Cevc and Blume, 2004; Fesq *et al.*, 2003].

7.4.2. Antihistaminic

Usually delivered *per os* for eczema treatment, topical application of antihistaminic should enhance targetability, reduce side-effects with easy use. Ketotifen fumarate was used as an encapsulated model drug in skin delivery. Three carriers were compared: traditional liposomes, deformable liposomes (transfersomes) and ethosomes. They pointed out that traditional liposomes improved only total skin deposition of drug whereas both deformable liposomes and ethosomes improved also skin permeation [Elsayed *et al.*, 2007].

Hydroxyzine, a piperazine-class H_1-antihistaminic, was used for the treatment of histamine related skin disorder by conventional oral administration exhibiting dramatic side-effects [Elzainy *et al.*, 2003]. Topical application of first-generation H_1-antihistaminic in ointments and creams for treatment of symptoms of allergic skin disorders was recommended for many years; however, considerable systemic absorption was reported, leading potentially to systemic side effects.

Liposome encapsulation was suggested to overcome these limitations. Hydroxyzine loaded small unilamelar and multilamelar vesicles showed an excellent topical H_1-antihistaminic activity, and minimal systemic exposure [Elzainy *et al.*, 2003].

7.4.3. Triamcinolone Acetonide

Skin blanching and clinical efficacy of a liposomally entrapped triamcinolone acetonide cream was compared with that of the conventional triamcinolone cream in healthy human volunteers and eczema patients, respectively. Both creams showed equal efficacy in eczema patients. A significant reduction in the skin blanching response with the liposomal triamcinolone cream as compared to the conventional triamcinolone cream suggested a decrease in the systemic absorption of the corticosteroid [Rao *et al.*, 1994].

7.5. Mycosis

The main target in fungal treatment is ergosterol involved in the formation of the fungal cell membrane. Drugs can act by inhibiting enzymes involved in ergosterol synthetis (e.g., imidazole, triazole and alkylamines derivatives), or bind directly to the preformed sterol (i.e. polyene derivatives) leading to a more crystalline phase of fungal cell membrane and so lysis. A third type of treatment point glucan from cell wall, (e.g., echinocandins derivatives) which inhibit its formation and lead to desorganize cell wall, incompatible with fungi cell survival.

Clotrimazole is an imidazole antifungal drug active on *Candida albicans*, *Epidermophyton, Malassezia furfur, Microsporum, Trichophyton* and *Pityrosporum orbiculare*. Clotrimazole has been used as a highly lipophilic model drug in SLN and NLC [Souto *et al.*, 2004]. This study was principally focused on physical stability of these particles, as well as the entrapment efficiency of this lipophilic drug and its in vitro release profile. Particles sizes remained unchanged among time of storage but NLC showed a higher entrapment efficiency due to their liquid parts, a faster release profile in comparison to SLN with the same lipid concentration [Souto *et al.*, 2004].

7.6. Herpes

Herpes simplex virus 1 and 2 (HSV-1 and HSV-2) are two strains of the Herpes virus family, *Herpesviridae*, which cause infections in humans. After an initial, or *primary*, infection, HSV establishes *latency*, during which the virus is present in the cell bodies of nerves which innervate the area of original outbreak. During *reactivation*, virus is produced in the cell and transported outwardly via the nerve cell's axon to the skin. The ability of Herpes virus to establish latency leads to the chronic nature of Herpes infection. After the initial infection subsides, Herpes symptoms may periodically recur in the form of *outbreaks* of herpetic sores near the original infection site.

Herpes infections are marked by painful, watery blisters in the skin or mucous membranes (such as the mouth or lips) or on the genitals. Treatments involve antiviral like ibacitabine, or more often acyclovir which product a selective inhibition on viral DNA polymerase following activation by viral thymidine kinase.

Liposomal acyclovir delivery system was proposed [Chetoni *et al.*, 2004; Pavelic *et al.*, 2005] to treat vaginal and corneal infections. Positively charged liposomes bind intimately to the negatively charged corneal surface, leading to an increased residence time and a superior drug bioavailability [Chetoni *et al.*, 2004; Law *et al.*, 2000]. Chetoni reported also a superior bioavailability of acyclovir from positively charged liposomal formulation, when compared on the same dose with the commercially available ointment.

Herpes labialis is usually treated with commercial 5% acyclovir cream. Horwitz [Horwitz *et al.*, 1999], showed a significant and earlier improvement of labial lesions treated by ethosomal formulation as compared to the cream form and drug-free vehicle.

8. Conclusion

Nanoencapsulation is technology presenting obvious and rising advantages for the treatment of skin diseases. Nanocarriers (e.g. liposomes, niosomes, transfersomes, SLN, nanoparticles and microemulsions) are well defined and characterized systems. Therefore, the encapsulation of drug within these carriers for topical application should be considered in term of skin tolerance, cutaneous bioavailability and industrial scale-up.

Biophysically, nanocarriers open a window for the delivery of molecules considered as poor candidates in topical formulation (e.g., high molecular mass, high hydrophily) by enhancing their penetration and permeation through the skin. Furthermore, the use of drugs presenting a weak therapeutic index may be re-considered by the drug loading in colloids which reduces systemic side-effects. Technologically, the straightforward production of nanocarriers avoiding organic solvents should be a relevant criterion for industrial scale-up (e.g., SLN and microemulsions) conditioning the becoming of topical formulations for the 21st century's dermatological practices.

References

Agache, P. (2004). The human skin: An overview. In: P. Agache and P. Humbert (Editors), *Measuring the skin*, Springer-Verlag, Berlin Heidelberg, pp. 3-5.

Agarwal, R., Katare, O. P. and Vyas, S. P. (2001). Preparation and in vitro evaluation of liposomal/niosomal delivery systems for antipsoriatic drug dithranol. *Int. J. Pharm.*, 228, pp. 43-52.

Agarwal, R., Saraswat, A., Kaur, I., Katare, O. P. and Kumar, B. (2002). A novelliposomal formulation of dithranol for psoriasis: preliminary results. *J. Dermatol.*, 29, pp. 529-32.

Alberti, I., Kalia, Y. N., Naik, A., Bonny, J. D. and Guy, R. H. (2001). In vivo assessment of enhanced topical delivery of terbinafine to human stratum corneum. *J. Control. Rel.*, 71, pp. 319-327.

Alsarra, I. A., Bosela, A. A., Ahmed, S. M. and Mahrous, G. M. (2005). Proniosomes as a drug carrier for transdermal delivery of ketorolac. *Eur. J. Pharm. Biopharm.*, 59, pp. 485-490.

Alvarez-Figueroa, M. J. and Blanco-Mendez, J. (2001). Transdermal delivery of methotrexate: iontophoretic delivery from hydrogels and passive delivery from microemulsions. *Int. J. Pharm.*, 215, pp. 57-65.

Baroli, B., Lopez-Quintela, M. A., Delgado-Charro, M. B. and Fadda, A. M., Blanco-Mendez, J. (2000). Microemulsions for topical delivery of 8-methoxsalen. *J. Control. Rel.*, 69, pp. 209-218.

Beetge, E., du Plessis, J., Muller, D. G., Goosen, C. and van Rensburg, F. J. (2000). The influence of the physicochemical characteristics and pharmacokinetic properties of selected NSAID's on their transdermal absorption. *Int. J. Pharm.*, 193, pp. 261-264.

Brinon, L., Geiger, S., Alard, V., Doucet, J., Tranchant, J.-F. and Couarraze, G. (1999). Percutaneous absorption of sunscreens from liquid crystalline phases. *J. Control. Rel.*, 60, pp. 67-76.

Castelli, F., Puglia, C., Sarpietro, M. G., Rizza, L. and Bonina, F. (2005 a). Characterization of indomethacin-loaded lipid nanoparticles by differential scanning calorimetry. *Int. J. Pharm.*, 304, pp. 231-238.

Cevc, G. (2004). Lipid vesicles and other colloids as drug carriers on the skin. *Adv. Drug Deliv. Rev.*, 56, pp. 675-711.

Cevc, G., Blume, G. and Schatzlein, A. (1997). Transfersomes-mediated transepidermal delivery improves the regio-specificity and biological activity of corticosteroids in vivo. *J. Control. Rel.*, 45, pp. 211-226.

Cevc, G., Blume, G., Schatzlein, A., Gebauer, D. and Paul, A. (1996). The skin: a pathway for systemic treatment with patches and lipid-based agent carriers. *Adv. Drug Deliv. Rev.*, 18, pp. 349-378.

Cevc, G. and Blume, G. (1992). Lipid vesicles penetrate into intact skin owing to the transdermal osmotic gradients and hydration force. *Biochim. Biophys. Acta*, 1104, pp. 226-232.

Cevc, G. and Gebauer, D. (2003). Hydration-driven transport of deformable lipid vesicles through fine pores and the skin barrier, *Biophys. J.*, 84, pp. 1010-1024.

Cevc, G., Schatzlein, A. and Blume, G. (1995). Transdermal drug carriers: Basic properties, optimization and transfer efficiency in the case of epicutaneously applied peptides. *J. Control. Rel.*, 36, pp. 3-16.

Cevc, G., Schatzlein, A. and Richardsen, H. (2002). Ultradeformable lipid vesicles can penetrate the skin and other semi-permeable barriers unfragmented. Evidence from double label CLSM experiments and direct size measurements. *Biochim. Biophys. Acta*, 1564, pp. 21-30.

Chen, H., Chang, X., Du, D., Liu, W., Liu, J., Weng, T., Yang, Y., Xu, H. and Yang, X. (2006 b). Podophyllotoxin-loaded solid lipid nanoparticles for epidermal targeting. *J. Control. Rel.*, 110, pp. 296-306.

Chetoni, P., Rossi, S., Burgalassi, S., Monti, D., Mariotti, S. and Saettone, M. F. (2004). Comparison of Liposome-Encapsulated Acyclovir with Acyclovir Ointment: Ocular Pharmacokinetics in Rabbits. *J. Ocular Pharmacol. Therap.*, 20, pp. 169-177.

Danielsson, I. and Lindman, B. (1981). The definition of microemulsion. *Colloids and Surfaces*, 3, pp. 391-392.

Date, A. A., Naik, B. and Nagarsenker, M. S. (2006). Novel drug delivery systems: Potential in improving topical delivery of antiacne agents. *Skin Pharmacol. Physiol.*, 19, pp. 2-16.

Dayan, N. and Touitou, E. (2000). Carriers for skin delivery of trihexyphenidyl HCl: ethosomes vs. liposomes. *Biomaterials*, 21(18), pp. 1879-1885.

de Rie, M. A., Goedkoop, A. Y. and Bos, J. D. (2004). Overview of psoriasis. *Dermatol. Ther.*, 17(5), pp. 341-349.

Degim, I. T. (2006). New tools and approaches for predicting skin permeability. *Drug Discovery Today*, 11(11-12), pp. 517-523.

Delneuville, I., Dechesne, J. P. and Delattre, L. (1998). Preparation and study of the characteristics of dithranol:polyvinylpyrrolidone coevaporates, *Int. J. Pharm.*, 168, pp. 109-118.

Dubey, V., Mishra, D., Jain, N. K. (2007). Melatonin loaded ethanolic liposomes: Physicochemical characterization and enhanced transdermal delivery, *Eur. J. Pharm. Biopharm.,* in press.

Elsayed, M. M. A., Abdallah, O. Y., Naggar, V. F. and Khalafallah, N. M. (2007). Deformable liposomes and ethosomes: Mechanism of enhanced skin delivery. *Int. J. Pharm.*, 322, pp. 60-66.

Elsayed, M. M. A., Abdallah, O. Y., Naggar, V. F. and Khalafallah, N. M. Lipid vesicles for skin delivery of drugs: Reviewing three decades of research. *Int. J. Pharm.*, in press.

Elzainy, A. A., Gu, X., Simons, F. E. and Simons K. J. (2003). Hydroxyzine from topical phospholipid liposomal formulations: evaluation of peripheral antihistaminic activity and systemic absorption in a rabbit model. *AAPS PharmSci*, 5, pp. E28.

Esposito, E., Cortesi, R., Drechsler, M., Paccamiccio, L., Mariani, P., Contado, C., Stellin, E., Menegatti, E., Bonina, F. and Puglia, C. (2005). Cubosome dispersions as delivery systems for percutaneous administration of indomethacin, *Pharm. Res.*, 22, pp. 2163-2173.

Fang, J.-Y., Hong, C.-T., Chiu, W.-T. and Wang, Y.-Y. (2001a). Effect of liposomes and niosomes on skin permeation of enoxacin. *Int. J. Pharm.*, 219, pp. 61-72.

Fesq, H., Lehmann, J., Kontny, A., Erdmann, I., Theiling, K., Rother, M., Ring, J., Cevc, G. and Abeck, D. (2003). Improved risk-benefit ratio for topical triamcinolone acetonide in Transfersome in comparison with equipotent cream and ointment: a randomized controlled trial. *Br. J. Dermatol.*, 149, pp. 611-619.

Fluhr, J. W., Barsom, O., Gehring, W. and Gloor, M. (1999). Antibacterial efficacy of benzoyl peroxide in phospholipid liposomes. A vehicle-controlled, comparative study in patients with papulopustular acne. *Dermatology*, 198, pp. 273-277.

Fogh, K. and Kragballe, K. (2004). New vitamin D analogs in psoriasis, *Curr. Drug Targets Inflamm. Allergy*, 3, pp. 199-204.

Foldvari, M., Gesztes, A. and Mezei, M. (1990). Dermal drug delivery by liposome encapsulation: clinical and electron microscopic studies, *J. Microencaps.*, 7, pp. 479-489.

Fresno Contreras, M. J., Jimenez Soriano, M. M. and Ramirez Dieguez, A. (2005). In vitro percutaneous absorption of all-trans retinoic acid applied in free form or encapsulated in stratum corneum lipid liposomes. *Int. J. Pharm.*, 297, pp. 134-145.

Fresta, M. and Puglisi, G. (1996). Application of liposomes as potential cutaneous drug delivery systems. In vitro and in vivo investigation with radioactively labelled vesicles. *J. Drug Target.*, 4(2), pp. 95-101.

Fresta, M. and Puglisi, G. (1997). Corticosteroid dermal delivery with skin-lipid liposomes. *J. Control. Rel.*, 44, pp. 141-151.

Geinoz, S., Guy, R. H., Testa, B. and Carrupt, P.-A. (2004). Quantitative structure-permeation relationships (QSPeRs) to predict skin permeation: a critical evaluation. *Pharm. Res.*, 21, pp. 83-92.

Godin, B. and Touitou, E. (2004). Mechanism of bacitracin permeation enhancement through the skin and cellular membranes from an ethosomal carrier. *J. Control. Rel.*, 94(2-3), pp. 365-379.

Gomes, A. J., Faustino, A. S., Lunardi, C. N., Lunardi, L. O. and Machado, A. E. H. (2007). Evaluation of nanoparticles loaded with benzopsoralen in rat peritoneal exudate cells. *Int. J. Pharm.*, 332, 153-160.

Gopinath, D., Ravi, D., Rao, B. R., Apte, S. S., Renuka, D. and Rambhau, D. (2004). Ascorbyl palmitate vesicles (Aspasomes): formation, characterization and applications. *Int. J. Pharm.*, 271, pp. 95-113.

Gruber, D. M., Sator, M. O., Joura, E. A., Kokoschka, E. M., Heinze, G., Huber, J. C. (1998). Topical cyproterone acetate treatment in women with acne: a placebo-controlled trial. *Arch. Dermatol.*, 134, pp. 459-463.

Hadgraft, J. (2004). Skin deep. *Eur. J. Pharm. Biopharm.*, 58, pp. 291-299.

Heurtault, B., Saulnier, P., Pech, B., Proust, J.-E. and Benoit, J.-P. (2003). Physico-chemical stability of colloidal lipid particles. *Biomaterials*, 24, pp. 4283-4300.

Hill, J. R., Wertz, P. W. (2003). Molecular models of the intercellular lipid lamellae from epidermal stratum corneum. *Biochim. Biophys. Acta*, 1616, pp. 121-126.

Hiruta, Y., Hattori, Y., Kawano, K., Obata, Y. and Maitani, Y. (2006). Novel ultra-deformable vesicles entrapped with bleomycin and enhanced to penetrate rat skin. *J. Control. Rel.*, 113, pp. 146-154.

Honzak, L. and Sentjurc, M. (2000). Development of liposome encapsulated clindamycin for treatment of acne vulgaris. *Pflugers Arch.*, 440(5 Suppl), pp. R44-45.

Horwitz, E., Pisanty, S., Czerninski, R., Helser, M., Eliav, E. and Touitou, E. (1999). A clinical evaluation of a novel liposomal carrier for acyclovir in the topical treatment of recurrent herpes labialis. *Oral Surg. Oral Med. Oral Pathol. Oral Radiol. Endod.*, 87, pp. 700-705.

Hou, D., Xie, C., Huang, K. and Zhu, C. (2003). The production and characteristics of solid lipid nanoparticles (SLNs). *Biomaterials*, 24, pp. 1781-1785.

Hu, F. Q., Jiang, S. P., Du, Y. Z., Yuan, H., Ye, Y. Q. and Zeng, S. (2006). Preparation and characteristics of monostearin nanostructured lipid carriers. *Int. J. Pharm.*, 314, pp. 83-89.

Hu, F.-Q., Jiang, S.-P., Du, Y.-Z., Yuan, H., Ye, Y.-Q. and Zeng, S. (2005). Preparation and characterization of stearic acid nanostructured lipid carriers by solvent diffusion method in an aqueous system. *Colloids and Surfaces B: Biointerfaces*, 45, pp. 167-173.

Iraji, F., Momeni, A., Naji, S. M. and Siadat, A. H. (2006). The efficacy of topical cyproterone acetate alcohol lotion versus placebo in the treatment of the mild to moderate acne vulgaris: a double blind study. *Dermatol. Online J.*, 12, pp. 26.

Jenning, V., Gysler, A., Schäfer-Korting, M. and Gohla, S. H. (2000a). Vitamin A loaded solid lipid nanoparticles for topical use: occlusive properties and drug targeting to the upper skin, Eur. J. Pharm. Biopharm., 49, pp. 211-218.

Jenning, V., Schäfer-Korting, M. and Gohla, S. (2000b). Vitamin A-loaded solid lipid nanoparticles for topical use: drug release properties. *J. Control. Rel.*, 66, pp. 115-126.

Jores, K., Mehnert, W., Drechsler, M., Bunjes, H., Johann, C. and Mader, K. (2004). Investigations on the structure of solid lipid nanoparticles and oil-loaded solid lipid nanoparticles by photon correlation spectroscopy, field-flow fractionation and transmission electron microscopy. *J. Control. Rel.*, 95, pp. 217-227.

Kalia, Y. N., Alberti, I., Naik, A. and Guy, R. H. (2001). Assessment of topical bioavailability in vivo: the importance of stratum corneum thickness, *Skin Pharmacol. Appl. Skin Physiol.*, 14, pp. 82-86.

Karzel, K. and Liedtke, R. K. (1989). Mechanisms of transcutaneous absorption. Pharmacologic and biochemical aspects. *Arzneimittel-Forschung*, 39, pp. 1487-1491.

Kemken, J., Ziegler, A. and Muller, B. W. (1992). Influence of supersaturation on the pharmacodynamic effect of bupranolol after dermal administration using microemulsions as vehicle. *Pharm. Res.*, 9, pp. 554-8.

Kitagawa, S. and Kasamaki, M. (2006). Enhanced delivery of retinoic acid to skin by cationic liposomes. *Chem. Pharm. Bull.*, 54, pp. 242-244.

Kogan, A. and Garti, N. (2006). Microemulsions as transdermal drug delivery vehicles. *Adv. Colloid Interface Sci.*, 123-126, pp. 369-385.

Korbel, J. N., Sebok, B., Kerenyi, M. and Mahrle, G. (2001). Enhancement of the antiparakeratotic potency of calcitriol and tacalcitol in liposomal preparations in the mouse tail test. *Skin Pharmacol. Appl. Skin Physiol.*, 14, pp. 291-5.

Korting, H. C., Zienicke, H., Schafer-Korting, M. and Braun-Falco, O. (1990). Liposome encapsulation improves efficacy of betamethasone dipropionate in atopic eczema but not in psoriasis vulgaris. *Eur. J. Clin. Pharmacol.*, 39, pp. 349-351.

Kreilgaard, M. (2002). Influence of microemulsions on cutaneous drug delivery. *Adv. Drug Deliv. Rev.*, 54, pp. S77-98.

Lademann, J., Richter, H., Teichmann, A., Otberg, N., Blume-Peytavi, U., Luengo, J., Weiss, B., Schaefer, U. F., Lehr, C.-M., Wepf, R. and Sterry, W. (2007). Nanoparticles - An efficient carrier for drug delivery into the hair follicles. *Eur. J. Pharm. Biopharm.,* 66:159-164.

Law, S. L., Huang, K. J. and Chiang, C. H. (2000). Acyclovir-containing liposomes for potential ocular delivery. Corneal penetration and absorption. *J Control. Rel.*, 63, pp. 135-140.

Lboutounne, H., Faivre, V., Falson, F. and Pirot, F. (2004 a). Characterization of transport of chlorhexidine-loaded nanocapsules through hairless and Wistar rat skin. *Skin Pharmacol. Appl. Skin Physiol.*, 17, pp. 176-182.

Lboutounne, H., Guillaume, Y. C., Michel, L., Makki, S., Humbert, P. and Millet, J. (2004 b). Study and Development of Encapsulated Forms of 4,5',8-Trimethylpsoralen for Topical Drug Delivery, *Drug Dev. Res.*, 61, pp. 86-94.

Lebwohl, M., Ting, P. T. and Koo, J. Y. (2005). Psoriasis treatment: traditional therapy, *Ann. Rheum. Dis.*, 64 Suppl 2, pp. 83-86.

Liu, J., Hu, W., Chen, H., Ni, Q., Xu, H. and Yang, X. (2007). Isotretinoin-loaded solid lipid nanoparticles with skin targeting for topical delivery. *Int. J. Pharm.*, 328, pp. 191-195.

Lodzki, M., Godin, B., Rakou, L., Mechoulam, R., Gallily, R. and Touitou, E. (2003). Cannabidiol--transdermal delivery and anti-inflammatory effect in a murine model. *J. Control. Rel.*, 93, pp. 377-387.

Luengo, J., Weiss, B., Schneider, M., Ehlers, A., Stracke, F., Konig, K., Kostka, K. H., Lehr, C. M. and Schaefer, U. F. (2006). Influence of nanoencapsulation on human skin transport of flufenamic acid. *Skin Pharmacol. Physiol.*, 19, pp. 190-7.

Maia, C. S., Mehnert, W. and Schafer-Korting, M. (2000). Solid lipid nanoparticles as drug carriers for topical glucocorticoids. *Int. J. Pharm.*, 196, pp. 165-167.

Maia, C. S., Mehnert, W., Schaller, M., Korting, H. C., Gysler, A., Haberland, A. and Schäfer-Korting, M. (2002). Drug targeting by solid lipid nanoparticles for dermal use. *J. Drug Target.*, 10, pp. 489-495.

Manconi, M., Sinico, C., Valenti, D., Lai, F. and Fadda, A. M. (2006). Niosomes as carriers for tretinoin: III. A study into the in vitro cutaneous delivery of vesicle-incorporated tretinoin. *Int. J. Pharm.*, 311, pp. 11-19.

Manconi, M., Sinico, C., Valenti, D., Loy, G. and Fadda, A. M. (2002). Niosomes as carriers for tretinoin. I. Preparation and properties. *Int. J. Pharm.*, 234, pp. 237-248.

Manconi, M., Valenti, D., Sinico, C., Lai, F., Loy, G., and Fadda, A. M. (2003). Niosomes as carriers for tretinoin. II. Influence of vesicular incorporation on tretinoin photostability. *Int. J. Pharm.*, 260, pp. 261-72.

Mehnert, W. and Mader, K. (2001). Solid lipid nanoparticles: Production, characterization and applications. *Adv. Drug Deliv. Rev.*, 47, pp. 165-196.

Meidan, V. M., Bonner, M. C. and Michniak, B. B. (2005). Transfollicular drug delivery - Is it a reality? *Int. J. Pharm.*, 306, pp. 1-14.

Merz, K. and Sternberg, B. (1994). Incorporation of vitamin D3-derivatives in liposomes of different lipid types. *J. Drug Target.*, 2, pp. 411-7.

Moghimi, H. R., Williams, A. C. and Barry, B. W. (1996). A lamellar matrix model for stratum corneum intercellular lipids. II. Effect of geometry of the stratum corneum on permeation of model drugs 5-fluorouracil and oestradiol. *Int. J. Pharm.*, 131, pp. 117-129.

Müller, R. H., Mader, K. and Gohla, S. (2000). Solid lipid nanoparticles (SLN) for controlled drug delivery - a review of the state of the art. *Eur. J. Pharm. Biopharm.*, 50, pp. 161-177.

Mura, S., Pirot, F., Manconi, M., Falson, F. and Fadda, A. M. (2007). Liposomes and niosomes as potential carriers for dermal delivery of minoxidil. *J. Drug Target.*, 15, pp. 101-108.

Norlen, L., Nicander, I., Lundh Rozell, B., Ollmar, S. and Forslind, B. (1999). Inter- and intra-individual differences in human stratum corneum lipid content related to physical parameters of skin barrier function in vivo. *J. Invest. Dermatol.*, 112, pp. 72-77.

Paolino, D., Lucania, G., Mardente, D., Alhaique, F. and Fresta, M. (2005). Ethosomes for skin delivery of ammonium glycyrrhizinate: In vitro percutaneous permeation through human skin and in vivo anti-inflammatory activity on human volunteers. *J. Control. Rel.*, 106, pp. 99-110.

Patel, V. B., Misra, A. and Marfatia, Y. S. (2000). Topical liposomal gel of tretinoin for the treatment of acne: research and clinical implications. *Pharm. Dev. Technol.*, 5, pp. 455-64.

Patel, V. B., Misra, A. and Marfatia, Y. S. (2001a). Clinical assessment of the combination therapy with liposomal gels of tretinoin and benzoyl peroxide in acne. *AAPS PharmSciTech*, 2, pp. E-TN4.

Patel, V. B., Misra, A. N. and Marfatia, Y. S. (2001b). Preparation and comparative clinical evaluation of liposomal gel of benzoyl peroxide for acne. *Drug Dev. Ind. Pharm.*, 27(8), pp. 863-869.

Pathak, M. A. and Fitzpatrick, T. B. (1992). The evolution of photochemotherapy with psoralens and UVA (PUVA): 2000 BC to 1992 AD. *J. Photochem. Photobiol. B.*, 14(1-2), pp. 3-22.

Pavelic, Z., Skalko-Basnet, N., Filipovic-Grcic, J., Martinac, A. and Jalsenjak, I. (2005). Development and in vitro evaluation of a liposomal vaginal delivery system for acyclovir. *J. Control. Rel.*, 106, pp. 34-43.

Pirot, F., Berardesca, E., Kalia, Y. N., Singh, M., Maibach, H. I. and Guy, R. H. (1998). Stratum corneum thickness and apparent water diffusivity: facile and noninvasive quantitation in vivo, *Pharm. Res.*, 15, pp. 492-494.

Pirot, F. and Falson, F. (2004). Skin barrier function. In: P. Agache and P. Humbert (Editors), Measuring the skin (Springer-Verlag, Berlin Heidelberg), pp. 513-524.

Potapenko, A. Y. and Kyagova, A. A. (1998). The application of antioxidants in investigations and optimization of photochemotherapy. *Membr. Cell Biol.*, 12, pp. 269-278.

Prufer, K. and Jirikowski, G. F. (1996). Liposomal incorporation changes the effect of 1.25-dihydroxyvitamin D3 on the phospholipase C signal transduction pathway and the eicosanoid cascade on keratinocytes in vitro. *Biochem. Pharmacol.*, 51, pp. 247-252.

Rao, G. U., Murthy, R. S. R., Marfaita, Y. S. and Shankar, R. (1994). Influence of liposomal drug entrapment on the performance of corticosteroid creams. *Indian J. Dermatol.*, 60, pp. 5-7.

Rolland, A., (1993). Particulate carriers in dermal and transdermal delivery: myth or reality? In: A. Rolland (Editor), *Pharmaceutical particulate carriers: therapeutic applications* (Marcel Dekker, New York), pp. 367-421.

Saraswat, A., Agarwal, R., Kaur, I., Katare, O. P. and Kumar, B. (2002). Fabric-staining properties and washability of a novel liposomal dithranol formulation. *J. Dermatolog. Treat.*, 13, pp. 119-122.

Saupe, A., Gordon, K. C. and Rades, T. (2006). Structural investigations on nanoemulsions, solid lipid nanoparticles and nanostructured lipid carriers by cryo-field emission scanning electron microscopy and Raman spectroscopy. *Int. J. Pharm.*, 314, pp. 56-62.

Schafer-Korting, M., Korting, H. C. and Ponce-Poschl, E. (1994). Liposomal tretinoin for uncomplicated acne vulgaris. *Clin. Investig.*, 72, pp. 1086-1091.

Shalita, A. R. (2004). Acne: Clinical presentations. *Clinics in Dermatology*, 22, pp. 385-386.

Simeonova, M., Velichkova, R., Ivanova, G., Enchev, V. and Abrahams, I. (2003). Poly(butylcyanoacrylate) nanoparticles for topical delivery of 5-fluorouracil. *Int. J. Pharm.*, 263, pp. 133-140.

Sinico, C., Manconi, M., Peppi, M., Lai, F., Valenti, D. and Fadda, A. M. (2005). Liposomes as carriers for dermal delivery of tretinoin: in vitro evaluation of drug permeation and vesicle-skin interaction. *J. Control. Rel.*, 103, pp. 123-136.

Skalko, N., Cajkovac, M. and Jalsenjak, I. (1992). Liposomes with clindamycin hydrochloride in the therapy of acne vulgaris, *Int. J. Pharm.*, 85, pp. 97-101.

Souto, E. B., Wissing, S. A., Barbosa, C. M. and Muller, R. H. (2004). Development of a controlled release formulation based on SLN and NLC for topical clotrimazole delivery. *Int. J. Pharm.*, 278, pp. 71-77.

Stecova, J., Mehnert, W., Blaschke, T., Kleuser, B., Sivaramakrishnan, R., Zouboulis, C. C., Seltmann, H., Korting, H. C., Kramer, K. D. and Schafer-Korting, M. (2007). Cyproterone acetate loading to lipid nanoparticles for topical acne treatment: particle characterisation and skin uptake. *Pharm. Res.*, 24, pp. 991-1000.

Stinchcomb, A. L., Pirot, F., Touraille, G. D., Bunge, A. L. and Guy, R. H. (1999). Chemical uptake into human stratum corneum in vivo from volatile and non-volatile solvents. *Pharm. Res.*, 16(8), pp. 1288-1293.

Surber, C. and Smith, E. W. (2005). The mystical effects of dermatological vehicles. *Dermatology*, 210(2), pp. 157-168.

Tabbakhian, M., Tavakoli, N., Jaafari, M. R. and Daneshamouz, S. (2006). Enhancement of follicular delivery of finasteride by liposomes and niosomes: 1. In vitro permeation and in vivo deposition studies using hamster flank and ear models. *Int. J. Pharm.*, 323, pp. 1-10.

Touitou, E., Dayan, N., Bergelson, L., Godin, B. and Eliaz, M. (2000). Ethosomes -- novel vesicular carriers for enhanced delivery: characterization and skin penetration properties. *J. Control. Rel.*, 65, pp. 403-418.

Touitou, E., Godin, B., Dayan, N., Weiss, C., Piliponsky, A. and Levi-Schaffer, F. (2001). Intracellular delivery mediated by an ethosomal carrier. *Biomaterials*, 22, pp. 3053-3059.

Trotta, M., Peira, E., Carlotti, M. E., and Gallarate, M. (2004). Deformable liposomes for dermal administration of methotrexate. *Int. J. Pharm.*, 270, pp. 119-25.

Trotta, M., Ugazio, E., Peira, E. and Pulitano, C. (2003). Influence of ion pairing on topical delivery of retinoic acid from microemulsions. *J. Control. Rel.*, 86, pp. 315-321.

Tschan, T., Steffen, H. and Supersaxo, A. (1997). Sebaceous-gland deposition of isotretinoin after topical application: an in vitro study using human facial skin. *Skin Pharmacol.*, 10, pp. 126-34.

van De Kerkhof, P. C. (2001). New developments in the treatment of psoriasis. *Skin Pharmacol. Appl. Skin Physiol.*, 14, pp. 129-35.

Verma, D. D. and Fahr, A. (2004). Synergistic penetration enhancement effect of ethanol and phospholipids on the topical delivery of cyclosporin A. *J. Control. Rel.*, 97, pp. 55-66.

Verma, D. D., Verma, S., Blume, G. and Fahr, A. (2003). Particle size of liposomes influences dermal delivery of substances into skin. *Int. J. Pharm.*, 258, pp. 141-151.

Vogt, A., Combadiere, B., Hadam, S., Stieler, K. M., Lademann, J., Schaefer, H., Autran, B., Sterry, W. and Blume-Peytavi, U. (2006). 40 nm, but not 750 or 1,500 nm, nanoparticles enter epidermal CD1a+ cells after transcutaneous application on human skin. *J. Invest. Dermatol.*, 126, pp. 1316-1322.

Vora, B., Khopade, A. J. and Jain, N. K. (1998). Proniosome based transdermal delivery of levonorgestrel for effective contraception. *J. Control. Rel.*, 54, pp. 149-165.

Weyenberg, W., Filev, P., Van den Plas, D., Vandervoort, J., De Smet, K., Sollie, P. and Ludwig, A. (2007). Cytotoxicity of submicron emulsions and solid lipid nanoparticles for dermal application. *Int. J. Pharm.*, 337, pp. 291-298.

Wissing, S. A. and Muller, R. H. (2002a). The influence of the crystallinity of lipid nanoparticles on their occlusive properties. *Int. J. Pharm.*, 242, pp. 377-379.

Wissing, S. A. and Muller, R. H. (2002 b). Solid lipid nanoparticles as carrier for sunscreens: in vitro release and in vivo skin penetration. *J. Control. Rel.*, 81, pp. 225-233.

Wissing, S. A. and Muller, R. H. (2003). Cosmetic applications for solid lipid nanoparticles (SLN). *Int. J. Pharm.*, 254, pp. 65-68.

Wong, T. W., Zhao, Y. L., Sen, A., Hui, S. W. (2005). Pilot study of topical delivery of methotrexate by electroporation. *Br. J. Dermatol.*, 152(3), pp. 524-530.

Yamaguchi, Y., Nagasawa, T., Nakamura, N., Takenaga, M., Mizoguchi, M., Kawai, S.-i., Mizushima, Y. and Igarashi, R. (2005). Successful treatment of photo-damaged skin of nano-scale atRA particles using a novel transdermal delivery. *J. Control. Rel.,* 104, pp. 29-40.

Chapter 6

NANOPARTICLES FOR ORAL VACCINATION

Juan M. Irache

Department of Pharmaceutical Technology
Faculty of Pharmacy, University of Navarra, Pamplona (Spain)

Hesham H. Salman

Department of Pharmaceutical Technology
Faculty of Pharmacy, University of Navarra, Pamplona (Spain)

Sara Gómez

Department of Pharmaceutical Technology
Faculty of Pharmacy, University of Navarra, Pamplona (Spain)

Carlos Gamazo

Department of Microbiology,
Faculty of Pharmacy, University of Navarra, Pamplona (Spain)

1. Introduction

Until recently, most vaccines presented antigens according to the same principals as they were a hundred years ago. These vaccines consist of the whole microorganism, alive or inactivated (bacterins), or attenuated toxins. Although these have proved successful, several drawbacks related with the safety (residual virulence, undesirable side effects) and the relatively poor efficacy against more challenging diseases (AIDS, malaria, tuberculosis) have limited their use.

In the past decades, several new approaches in vaccine development have shown to have significant advantages over traditional ones. The new generation of vaccines contains other defined antigens, called "subunit vaccines". They include the use of vaccines based in proteins, synthetic peptides, and plasmid DNA. Although they offer advantages such as reduced toxicity compared to traditional ones, they are poorly immunogenic when administered alone. For these reasons, accessory technologies are required to make these defined antigens more immunogenic, including new approaches for their optimal physical presentation to the antigen presenting cells (APCs) [O'Hagan and Valiante, 2003]. One of these technologies, and the most popular, is the use of adjuvants to improve the immunogenicity of the active molecule and confer to the host the required protection against infection.

This last fact is particularly challenging for mucosal delivery of antigens. Mucosal immunisation regimes that employ the oral route of delivery are often compromised by antigen degradation in the gastrointestinal tract. Moreover, tolerance or immunological unresponsiveness to orally delivered vaccine antigens may be also one of the main obstacles associated with this route of immunisation. On the other hand, although its commercial viability is yet to be proven, oral vaccination provides a cost-effective and convenient alternative to conventional vaccines. Moreover, an oral vaccine not only elicits a good immune response and may induce protection at the sites where pathogens often establish the infection, but it is also likely to eliminate the geographic limitations inherent in most vaccination programs.

2. Immunological Adjuvants

"Adjuvants" were originally described by Ramon as "substances used in combination with a specific antigen that produce more immunity than the antigen alone" [Ramon 1925; 1926]. Today, the term "adjuvant" (from the Latin word *adjuvare*, which means to help or to enhance) is used to identify any substance, combination of substances or strategies that augment specific immunity to an antigen as compared to that induced by the antigen or vaccine alone.

Another interesting point related to modern vaccination principles is the concept of immunomodulation. It includes the induction of specific antibodies of a desired isotype and IgG subclasses, the induction of selected T-helper cell responses as classified by the resulting cytokine production, the induction of cytotoxic T-cell responses, and the distribution of the immune response to several lymphatic sites, e.g. mucosal surfaces [McGhee *et al.*, 1992]. Any of these factors may be important to obtain protective immunity, the ultimate goal for a vaccine [Schijns, 2003].

Chemically, the adjuvants are a highly heterogeneous group of compounds with just only one thing in common: their ability to enhance the immune response. An effective adjuvant formulation provides the antigen with both an optimal physical presentation and a boost to create immune recognition and reaction [Marx *et al.*, 1993]. They are highly variable in terms of how they affect the immune system and what type of immunomodulation process they induce. In any case, specific antigen/adjuvant combinations preferentially induce type 1 (Th1) or type 2 (Th2) cytokine responses [Schijns, 2003]. The Th1 subset is characterised by the secretion of cytokines such as interleukin 2 (IL-2) and interferon-γ (IFN-γ), to assist in cell-mediated immune response. On the other hand, the Th2 subset assists preferentially in antibody immune responses after secreting cytokines including interleukin 4 (IL-4).

2.1. Adjuvant Classification

Despite the recognition of many different types of adjuvants, however, there is still little is known about their mode of action. In any case, it becomes apparent that there is now consensus on two major functional classes of adjuvants or immunopotentiators [Petrovsky and Aguilar, 2004].

The first category of adjuvants is formed by the antigen delivery systems (or facilitators of signal 1) which either modify the antigen such that prolonged residency is achieved or generate a particulate form that may be interpreted as a pathogen invasion signal by the host immune system [Banchereau and Steinman, 1998]. These adjuvants increase the antigen uptake by the antigen presenting cells [Ardavin, 2003]. Either they are directly engulfed by APCs or they form a depot of antigen that prolongs exposure and thus, the chance of antigen to be taken up by APCs. As a result, these adjuvants increase the provision of antigen-loaded APC's for cognate naive T cells, promoting up-regulating of costimulatory signals or MHC expression, inducing cytokine release and thus enhancement of the magnitude and duration of the immune response [Degen *et al.*, 2003].

In principle, mineral adjuvants and emulsion-like adjuvants can be ascribed to this first group of adjuvants. In both cases, it is well admitted that they form a depot of antigen in the site of injection.

Particulate delivery systems also belong to this first category of adjuvants [Ulmer, 2004]. APCs have evolved to engulf microorganisms, and thus, it is not surprising that particulate antigen with sizes in the range of pathogens act as adjuvant by direct targeting of antigen to these cells. In fact, the term of particulate delivery systems comprise any strategy addressed to endow an antigen with dimensions of a microorganism [Espuelas *et al.*, 2005]. They can be classified in two major groups, according to their lipidic or polymeric composition. Within the group of lipid-based particles we can distinguish liposomes, immunostimulatory complexes (ISCOMs) and virosomes (see Table 1). Among non-lipidic particles, the following systems can be ascribed: virus-like particles, microparticles and nanoparticles.

The second category of adjuvants or facilitators of signal 2 are characterised by their ability to directly activate immune cells during antigen recognition and presentation. Since the immune system evolved to free the host from noxious pathogens, these signal 2 facilitators, include pathogen invasion signals; in most cases conserved microbe-derived molecules or synthetic mimetics, or endogenous host-derived molecules produced by immune or damaged cells.

Table 1 Lipid-particulate adjuvants.

Adjuvant	**Observations**	**References**
ISCOM Iscomatrix	ISCOMs have a cage-like structure composed of Quil A, cholesterol, phospholipids, and protein ISCOMs are 30-40 nm in diameter, and are composed of 12-nm ring-like subunits ISCOM provides both antigen delivery and immunomodulatory properties Experimental ISCOM vaccine formulations have induced protective immunity to a number of microorganisms including retroviruses, parasites and bacteria Potential problems in the inclusion of antigen into the adjuvant	Morein *et al.*, 1984 Morein *et al.*, 1994 Morein *et al.*, 1995 Papadopoulou *et al.*, 1998 Pearse and Drane, 2005
Liposomes	They enhance both humoral and cellular immunity to protein and polysaccharide antigens. Humoral and cell-mediated immunity with protection against visceral leishmaniasis demonstrated in a hamster model Stability, manufacturing and quality problems seem to have been major factors behind the fact that till now no adjuvant based on liposomes has been registered for human use	Poulain-Godefroy *et al.*, 1998 Kersten *et al.*, 2004 Mazumdar *et al.*, 2005
Virosomes	Virosomes are liposomes containing functional viral membrane proteins Influenza virus virosome induce important cytotoxic T-cell responses They can be considered as delivery systems with intrinsic adjuvant activity. Virosomes have been approved in Europe for systemic vaccination against hepatitis A and influenza	Durrer *et al.*, 2003 Huckriede *et al.*, 2003 Moser *et al.*, 2003
Other lipidic particles	Cochleates	Perez *et al.*, 2004

To this heterogeneous group of adjuvants belongs inflammatory stimuli (frequently associated with vaccine administration), cytokines,

CD40L and "Pathogen-Associated Molecular Patterns" (PAMPs). PAMPs are molecular patterns shared by multiple microorganisms, not presented in mammalian cells, that activate immune cells through interaction with pattern-recognition receptors (PRR) [Sheikh *et al.*, 2000]. Lipopolysaccharide, murein, CpG motifs, flagellin and other main structural components of bacterial cells, such as cholera toxin B subunit and mycoplasma lipoproteins, are examples of PAMPs used empirically for a long time.

The functions of both adjuvant classes, however, are not mutually exclusive. Interestingly, a combination of both types of immunopotentiator functionalities may result in further optimisation of vaccine efficacy. Hence, a coupling of immunopotentiator on a delivery device may prolong its residence or facilitate the targeting of the antigen to relevant antigen presenting cells.

Nevertheless, in spite of the large list of compounds and strategies described as adjuvants, to date, the most popular adjuvant for human use remains to be the aluminium-based mineral salts (generically called alum) [Clements and Griffiths, 2002]. Although alum is able to induce a high T-helper 2 (Th2) immune response, it has little capacity to stimulate cellular Th1 immune response [Singh and O'Hagan, 1999], which is required for the protection against many intracellular pathogens such as, tuberculosis, malaria, leishmaniasis or HIV-hepatitis C virus (HCV). Besides, alum has the potential to cause severe local and systemic side effects including IgE mediated allergies [Gupta, 1998]. Other major disadvantages of aluminium salts are the unpredictable extent of adsorption of certain proteins, the inability to induce mucosal immunity, and the low stability of some dispersed antigens [Lindblad, 2004].

2.2. Criteria for Adjuvant Selection

The most important issue in adjuvant development is safety, which has restricted the development of new adjuvants since alum was first introduced more than 50 years ago [Levine *et al.*, 1997]. Although a number of different adjuvants have been investigated over the years and some of them successfully applied in veterinary [Singh and O'Hagan. 2003], these have failed in humans largely because of toxicity, stability, bioavailability and/or cost problems.

For standard prophylactic immunisation in healthy individuals, only adjuvants that induce minimal side effects will become acceptable. Additional issues that are important for adjuvant development include biodegradability, ease of manufacture, and applicability to a wide range of vaccines [Brayden, 2001].

Ideally, adjuvants should be stable with long shelf life, biodegradable, cheap to produce, not auto-immunogenic and promote the appropriate immune response (i.e. cellular or antibody immunity depending on requirements for protection) [Brayden, 2001; De Magistris, 2006]. In addition, there are marked differences in the efficacy of adjuvants depending on the administration route (i.e. between mucosal and parenteral routes).

3. Mucosal Vaccination

The mucosa is a door of entry for many pathogens. Although it is very difficult to generate mucosal antibodies through parenteral vaccination, it is possible to obtain mucosal as well as parenteral immunity by inoculating antigen by the mucosal route. For pathogens colonising mucosal surfaces or those having a mucosal route of entry, protection correlates well with a strong local mucosal response.

In this context, several adjuvants have been described and proposed for mucosal vaccination including monophosphoryl lipid A (MPL), CpG oligonucleotides, Cholera toxin and *Escherichia coli* heat-labile enterotoxin [Petrovsky and Aguilar, 2004]. Sometimes, these mucosal adjuvants need a parenteral prime-mucosal boost (i.e., MPL) or their combination with delivery systems such as microparticles and nanoparticles. However, the multiplicity of mucosal (oral) delivery systems and adjuvants tested experimentally has not yet yielded many effective candidates.

3.1. Oral Vaccination

The oral route is the ideal means of delivering prophylactic and therapeutic vaccines, offering significant advantages over parenteral delivery. According to the published data, the gut represents the largest mucosal organ. It daily produces 50-100 mg IgA per kg body weight versus 30 mg IgG per kg produced by the entire body [McGhee *et al.*, 1992]. In addition, unlike parenteral immunisation, oral delivery can induce mucosal immune responses. However, the oral route of vaccine delivery is the most difficult because of the numerous barriers posed by the gastrointestinal tract.

Thus, by the oral route, vaccine components such as proteins, polysaccharides and DNA are extremely labile and could be degraded and/or damaged during passage through the gastrointestinal tract or through mucosa with a particular chemical environment, if not adequately protected. In addition, the oral route raises problems of antigen dilution because of the large surface of the gastrointestinal tract and may require higher amount of antigen/adjuvant formulation. On the other hand, for oral vaccination, it appears to be particularly interesting to favour the uptake of antigen by M cells of Peyer's patches (PPs) [Brayden and Baird, 2004].

In summary, to facilitate effective oral immunisation the antigen must be protected and conducted to target specific sites of the epithelial cell surface to ensure their appropriate uptake. The selected adjuvant may also enhance the immune response by mechanisms already described: adsorption and depot effect, cytokine induction, complement activation, recruiting of different cell populations, the delivery to different APCs, the regulation of the expression via MHC class I or class II and the stimulation of the production of different subtypes of antibodies [Smith *et al.*, 2004].

3.2. Gut-associated Immune System

The mucosa of the gastrointestinal tract is covered by epithelial cells (mainly enterocytes and globet cells) that form a protective barrier. Epithelial cells regulate the flow of water and nutrients and control the access of antigens and pathogens through the intercellular tight junctions that restrict the passage from the lumen to the submucosa [Miyoshi and Takai, 2005].

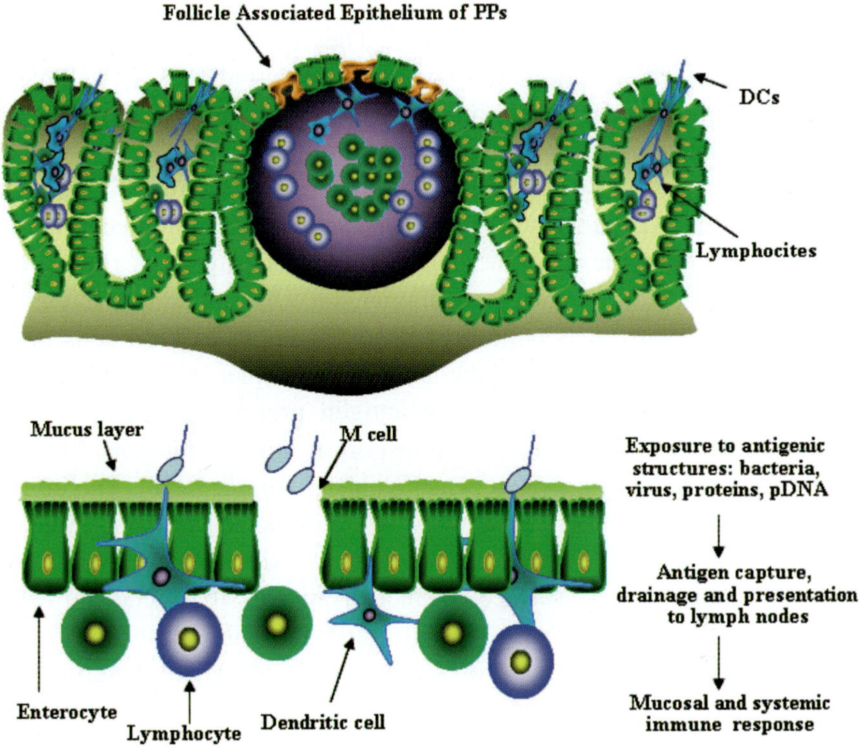

Fig. 1 Schematic representation of the gut-associated lymphoid tissue (GALT) and their role in the generation of an immune response. PPs: Peyer's patches; DCs: dendritic cells.

Interspersed among these epithelial cells are the components of the gut-associated lymphoid tissues (GALT). The GALT is represented by the PPs, the appendix and small solitary lymphoid nodules. All of these tissues serve as the mucosal inductive sites for the gastrointestinal tract [Ijima *et al.*, 2001]. In addition, a recent study has provided direct evidence that isolated lymphoid follicles (ILF) are equipped with immunological characteristics that are in some respects similar to those of PPs [Hamada *et al.*, 2002] and thus should be considered as a part of GALT. Figure 1 shows a schematic representation of gut associated lymphoid tissue and their role in the generation of immune responses.

The surface of PPs is covered by a unique epithelial layer known as follicle-associated epithelium (FAE). The FAE is enriched with specialised antigen-sampling cells known as microfold cells (M-cells), so named because of their unique characteristics of irregular and shortened microvilli [Owen and Jones, 1974]. From an anatomical point of view, PPs are characterised by possessing a dome configuration, with the area just below the FAE known to be enriched with IgA-committed B cells, Th1 and Th2 lymphocytes, macrophages and dendritic cells (DCs), all necessary for the induction of antigen-specific immune responses [Neutra *et al.*, 1996]. It is believed that antigen uptake by M cells does not result in the degradation of the antigen, but rather in the delivery of the intact antigen to the underlying antigen presenting cells [Nagler-Anderson, 2001]. In addition to the transport of luminal antigens, M cells serve as a port of entry for pathogens [Weinstein *et al.*, 1998]. Therefore, M cells are considered as a "gateway" to the mucosal immune system. However, this is not the only mechanism by which antigens may be captured at the mucosal surface. In fact, it appears that dendritic cells may form tight junctions with normal epithelial cells [Rescigno *et al.*, 2001]. This fact allows these professional APCs to extend dendrite-like processes through the epithelial layer and to sample luminal antigens directly [Rescigno *et al.*, 2001].

After the uptake and delivery of antigen via M cells, the antigens are immediately processed and presented by APCs such as DCs [Kelsall and Strober, 1996]. At least three DCs subpopulations have been found and described within the gut: myeloid DCs, lymphoid DCs and double negative DCs [Iwasaki and Kelsall, 2000]. Lymphoid and double negative DCs are capable of inducing Th1 differentiation for the subsequent development of cell-mediated immunity, while myeloid DCs induce Th2 cells for the generation of IgA immune responses in mucosal effector sites. Lymphoid and double negative DCs are classified as being of the DC1 subtype, while myeloid DCs are classified as being of the DC2 subtype [Iwasaki and Kelsall, 2001]. Further, the myeloid DCs produce IL-10 and TGF-α following exposure to innocuous food antigens and may mediate the T cell differentiation of Th3 or T regulatory (Tr) cells for the induction of systemic unresponsiveness to orally administered antigen known as oral tolerance or mucosally induced tolerance [Iwasaki and Kelsall, 2000].

In addition to dendritic cells, there are other immunocompetent cells such as macrophages, B- and T-lymphocytes and plasma cells [Neutra *et al.*, 1996]. After presentation of processed antigen by the different subsets of DCs described above, these Th cells can become Th1 and/or Th2 type cells for the induction, regulation and secretion of antigen-specific responses [London *et al.*, 1987].

Antibodies produced at this level and secreted to the lumen are mainly dimeric immunologlobulins of the IgA isotype (called "secretory IgA" or sIgA) [Rojas and Apodaca, 2002]. Locally produced secretory IgA is considered to be among the most important protective humoral factors. This antibody constitutes over 80% of all antibodies produced in mucosa associated tissues [McGhee *et al.*, 1992]. These IgA neutralise pathogens and their products either before mucosa invasion or during the colonisation process [Fernandez *et al.*, 2003]. In addition, IgA that are present in the lamina propria can also capture antigens such as toxins and viruses and excrete them from the mucosal tissue by intracellular cycling.

All of these processes appear to be controlled by cytokines. In fact, Th1 and Th2 cells are both reciprocally regulated by the cytokines they secrete; Th1 cells, by IFN-γ, and Th2 cells, by IL-4 and IL-10, among others [Coffman *et al.*, 1991]. These cytokines play an important role in the maintenance of appropriate immunological homeostasis in mucosa-associated compartments. On the other hand, Th3 cells produce TGF-β which promotes IgA isotype switching and has suppressive properties for both Th1 and Th2 cells [Weiner, 2001].

Another interesting point related with mucosal vaccination concerns the ability of B- and T-lymphocytes, that have been primed in mucosal tissues, to recirculate through different mucosa and home specifically in mucosal tissues because of the expression of specific homing receptors [Williams, 2004]. This aspect of the mucosal immune system is very important for vaccination because the immunisation at one mucosal site allows the appearance of antigen-specific lymphocytes in distant mucosal districts [Kunkel and Butcher, 2003; Williams, 2004].

3.3. Advantages of Mucosal Vaccination

The characteristics of the immune responses induced through the mucosal route confer several advantages over parenteral vaccination ones. Firstly, mucosal immunisation elicits antigen specific IgA antibodies at the mucosal site in which infection can occurs. This local pathogen-specific response may be important to prevent the infectious diseases in the vaccine recipient [Rojas and Apodaca, 2002]. Secondly, because of the migration of lymphocytes, the immunisation at one mucosal site can induce specific responses at distant sites [Williams, 2004; Kunkel and Butcher, 2003]. Finally, in addition to IgA responses, mucosal vaccination induces systemic IgG responses that represent a further defence against invasion by microorganisms or their products [Lillard *et al.*, 1999; Mann *et al.*, 2006]. Furthermore, mucosal vaccination could be exploited for combating pathogens acquired through non-mucosal routes (i.e.: blood or skin). In addition to serum IgG and mucosal IgA antibodies, mucosal immunisation can stimulate cell mediated responses including helper CD4+ T cells and CD8+ cytotoxic T lymphocytes, the latter being important to combat intracellular pathogens [Murillo *et al.*, 2002]. Thus, mucosal vaccines have the potential to activate all the different arms of the immune system.

On the other hand, mucosal administration of vaccines also offers a number of important practical advantages. First of all, this way of administration is non-invasive and does not require the use of needles. This would increase vaccine compliance and would also avoid problems of blood transmissible infections in developing countries due to needle re-use [Levine *et al.*, 2003]. Moreover, mucosal vaccination is relatively easy and does not require expensive specialised personnel. Reduced adverse effects and the potential for frequent boosting may also represent further advantages over injectable vaccines. Finally, production of mucosal vaccines may be less expensive than injectable vaccines that require high standards of purity, in addition to sterility.

4. Nanoparticles as Oral Adjuvants

Intuitively, oral vaccination using antigens loaded or encapsulated in particles appears to have a sound scientific rationale based on the protection of an antigen from exposure to gastric acid, bile and pancreatic secretions. At the same time, advantage is taken of the inherent inclination of M cells to take up particulates as part of its duty as a sentinel in triggering of mucosal immunity against enteric pathogens [Krahenbuhl and Neutra, 2000]. In fact, it has been clearly demonstrated that nanoparticles can interact with different components of the mucosa (including Peyer's patches) [Desai *et al.*, 1996; Arbós *et al.*, 2003] and provide a depot-effect [Storni *et al.*, 2005]. In addition, the use of nanoparticles offers another advantage such as the possibility to load an immunoadjuvant (i.e., CTB or LPS) with the antigen in order to increase and induce a determined immune response [O'Hagan *et al.*, 1995].

4.1. Factors Influencing the Efficacy of Nanoparticles as Oral Adjuvant

For "conventional" or non-decorated nanoparticles, the physico-chemical properties of these carriers appear to be the major factors influencing their efficacy as adjuvants. Among other properties, the particle size and the surface characteristics of the carriers strongly modulate and determine the type and intensity of the immune response within the gastrointestinal tract. Based on data published by Eldridge *et al.* [1989], asserting that particles smaller than 5 μm in diameter were transported by M cells through the efferent lymphatic macrophages, numerous particulate systems were developed to reach the goal of oral immunisation. However, particles in the nanoscale size rather than microscale size are more adapted for cellular uptake in the GI tract [Shakweh *et al.*, 2004].

Another important fact is related with the nature of the polymer used for oral vaccination. This aspect determines the stability of the resulting particles, their interaction with components of the mucosa and the release profile of the loaded antigen. Finally, it is also interesting to consider the influence of the antigen dose and the elicited immune response. However, little information is available concerning the of how stable particles really need to be, of how the antigen has to be released from the nanoparticles in the gastrointestinal tract, or as to whether M cells really take up more than just a few particles *in vivo*.

It seems to be clear that nanoparticles are more adapted to reach the GALT than microparticles [Tabata *et al.*, 1996]. Jepson and co-workers [1993] quantified this phenomenon of M-cell uptake in rabbit in approximately 10^5 particles of about 500 nm per each lymphoid follicle dome in 45 min. However, the real influence of the particle size remains to be controversial. Thus, in a recent work using bovine serum albumin (BSA) as antigen model, a higher serum IgG antibody level and a similar IgG2a/IgG1 ratio have been observed working with 1 μm PLGA particles than with the 200 and 500 nm particles [Gutierro *et al.*, 2002a].

Probably because of their (pseudo)lipophilic character, these drug delivery systems have proven to be taken up by the gut PPs [Singh and O'Hagan, 2003]. In addition, it has been observed that these delivery systems can be taken up by antigen presenting cells such as dendritic cells in vitro and in vivo [Lutsiak, 2002; Newman *et al.*, 2002].

A great number of antigens has been successfully encapsulated in PLGA particles with a full maintenance of structural and antigenic integrity. In general, ovalbumin, peptides, bacterial toxoids, inactivated bacteria and, more recently, DNA plasmids entrapped in PLGA particles have demonstrated to significantly enhance the systemic and/or mucosal immune responses after oral administration [Maloy, 1994; Kim *et al.*, 1999]. Furthermore, PLGA microparticles have already been used in humans as controlled release drug delivery systems and for other biomedical purposes [Okada and Toguchi, 1995] and clinical trials have been performed with oral vaccines incorporated in PLGA microparticles [Katz *et al.*, 2003].

Chitosan may also have an immunomodulatory effect as it has been shown to stimulate production of cytokines from immune cells in vitro

[Otterlei *et al.*, 1994] and enhance a naturally Th2/Th3-biased microenvironment at the mucosal level in absence of antigen [Porporatto *et al.*, 2005].

Recently, the copolymers between methyl vinyl ether and maleic anhydride (PVM/MA, Gantrez®) have also been proposed as a material to prepare nanoparticles for oral antigen delivery [Arbos *et al.*, 2002]. These copolymers are widely employed as thickening and suspending agents, denture adhesives, and excipient for the preparation of transdermal patches.

Finally, solid lipid nanoparticles [Olbrich *et al.*, 2002] and nanoparticles made from cationic cross-linked polysaccharides [Peppoloni *et al.*, 2003] have also been proposed as antigen delivery systems for oral vaccination.

Other important factors affecting the ability of nanoparticles to elicit an adequate immune response may be the antigen loading and the antigen release profile. In fact a direct relationship between antigen dose and IgG response was put in evidence with BSA-loaded particles when administered by the oral route [Gutierro *et al.*, 2002a; 2002b]. Similarly, a mixture of PLGA 50:50 and 75:25 particles (1:1) was proposed as the most proper formulation to immunize by a mucosal route [Rosas *et al.*, 2001]. Last but not least, the modification of the surface properties of nanoparticles may also dramatically affect their stability, distribution within the gut and interaction with the GALT. Thus, the encapsulation of chitosan nanoparticles in lipidic vesicles has been proposed to make them acid resistant upon oral administration and to induce significative higher IgG and sIgA titers than unmodified chitosan nanoparticles [Jain *et al.*, 2006].

4.2. Immunization with Nanoparticles

Up to now, few studies have examined the capacity of biodegradable loaded-nanoparticles to induce immune and protective responses after oral administration (Table 2).

Oral administration of antigen-loaded nanoparticles leads to the capture of these particulate systems by GALT. This process has been considered as a critical step to present the antigen to the lymphoid cells. Antigen presentation mainly occurs in PP, lymphoid nodules enriched in

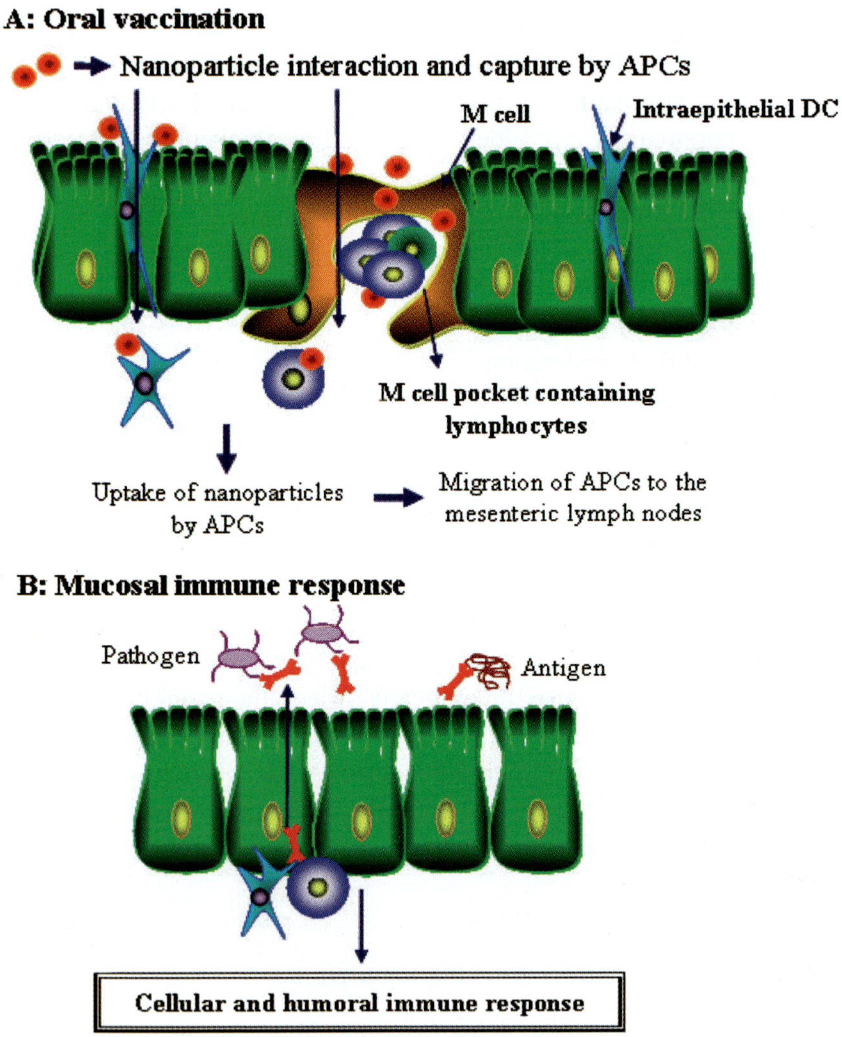

Fig. 2 Oral vaccination with antigen-loaded nanoparticles.

T and B lymphocytes, and specialised antigen presenting cells, M cells, which can transport antigen structures to T cells. Macromolecules, microorganisms and particles would be also taken up by DCs in the subepithelial dome region. These DCs may process and present antigen locally or migrate to mesenteric lymph nodes to interact with T cells. Figure 2 shows a schematic representation of the fate of antigen-loaded nanoparticles within the gut.

4.2.1. Antigen-loaded Nanoparticles

In mice, oral immunization with PLGA nanoparticles induced mucosal and systemic immunity to entrapped antigens. In this context, Kim *et al.* [1999] administrated PLGA nanoparticles containing *Helycobacter pylori* lysates to stimulate *H. pylori*-specific mucosal and systemic immune responses in Balb/c mice. They observed that these nanoparticles induced a strong specific mucosal IgA response as well as serum IgG1 and IgG2b responses, which were in all cases higher than for control animals (immunized with soluble antigens alone). In addition, these authors suggested that PLGA nanoparticles might be a safer adjuvant than cholera toxin.

More recently, Carcaboso *et al.* [2003] proposed the use of PLGA particles for oral vaccination against malaria using the synthetic peptide Spf66. In this work a mixture of PLGA 50:50 and 75:25 particles (ranging from 0.8 to 2 μm) were administered by the oral route, and when animals were boosted 3 weeks later significant systemic IgG antibody responses were elicited. These responses were comparable to those obtained with a traditional vaccine (alum triple shot) and superior to the aqueous vaccine given by the oral route [Carcaboso *et al.*, 2003].

In another study, it was shown that mucosal immunity towards Tat protein could be triggered in mice by the oral route when this protein is loaded in chitosan nanoparticles. Tat protein plays an essential role in HIV-1 replication and also participates in T-cell immunosuppression [Ensoli *et al.*, 1999; Le Buanec *et al.*, 2001]. In fact, sera from immunized mice with chitosan nanoparticles induced a cell-mediated immunity able to inhibit the activity of Tat protein, which is a prerequisite for the development of an anti-AIDS protective vaccine [Le Buanec *et al.*, 2001].

Nevertheless, despite the increased interest in the development of nanoparticles as oral adjuvants for antigen delivery, few studies address their potential efficacy against a challenge. Conway *et al.* [2001] have investigated the immunogenicity and protective efficacy of orally delivered *Bordetella pertussis* antigen entrapped in PLGA nanoparticles against a murine respiratory challenge model. Orally administered encapsulated antigens elicited not only mucosal responses but also systemic ones, leading to a protection against this pathogen. Specific IgA and IgG responses have been induced by oral immunization with the encapsulated antigen.

More recently, a hot saline extract of *Salmonella* Enteritidis was loaded in Gantrez nanoparticles (200 nm). In mice, three weeks after challenge with a lethal dose of this microorganism, the protection conferred by immunization was 80% and all the animals survived. By contrast, the control formulation conferred only 20% of protection and all the mice were died 6 days after the challenge [Gamazo *et al.*, 2004].

4.2.2. Nanoparticles for DNA Immunization

DNA plasmids can be encapsulated in nanoparticles with significant retention of biological function, and the oral administration of encapsulated DNA can elicit systemic and mucosal antibody responses to the encoded protein.

Jones *et al.* [1997] developed a method for encapsulation of plasmid DNA permitting its oral administration. In this work, plasmid DNA encoding insect luciferase was encapsulated in PLGA particles, ranging from 0.01 to 10 μm. The oral administration of this particles stimulated serum IgG, IgM, and IgA antibodies to luciferase. In addition, luciferase-specific IgA was also detected in stool samples, indicating a mucosal response [Jones *et al.*, 1997]. More recently, Bivas-Benita and collaborators [2003] compared the potential of chitosan nanoparticles (500 nm) loaded with *Toxoplasma gondii* GRA1 encoding DNA plasmid (pDNA) and chitosan microparticles loaded with recombinant GRA-1 protein to elicit GRA-1-specific immune responses after intragastric administration using different prime/ boost regimen. Boosting with GRA1 DNA vaccine resulted in high anti-GRA1 antibody levels, characterized by similar serum antibody titers of IgG2a and IgG1.

Table 2 Non-exhaustive list of oral immunization using antigen-loaded nanoparticles in mice.

Antigen	**Polymer**	**Observations**	**Refs**
BSA	PLGA	Higher serum IgG antibody levels and a similar IgG2a/IgG1 ratio with 1 μm particles than with the 200 and 500 nm particles	Gutierro *et al.*, 2002a; 2002b
Antigen from *Bordetella pertussis*	PLGA	Three doses conferred protection against *B. pertussis* challenge Induction of spleen cells (secretion of IL-5) and IgA in the lungs	Conway *et al.*,
H. pylori lysates	PLGA	Significantly specific IgA and IgG responses for at least 9 weeks	Kim *et al.*, 1999
Malaria peptide SPf66	PLGA	Systemic IgG antibody responses PLGA-vaccinated mice displayed a predominant Th1 response	Carcaboso *et al.*, 2003
Salmonella enteritidis extract (HE)	Gantrez	HE-NP elicited a Th1 response HE-NP displayed a higher protection than controls	Gamazo *et al.*, 2004
Rotavirus antigens DNA	PLGA	Intestinal IgA antibodies protected mice against rotavirus challenge	Herrmann *et al.*, 1999
DNA for *T. gondii* GRA1 protein	Chitosan	Boosting with GRA1 DNA vaccine resulted in high anti-GRA1 antibody levels, characterized by a mixed IgG2a/IgG1 ratio	Bivas-Benita *et al.*, 2003
DNA for a peanut allergen	Chitosan	Production of IgA, IgG2a and reduction in allergen-induced anaphylaxis	Roy *et al.*, 1999
House dust mite Der p1/ pDNA	Chitosan	Specific IgG2a and IgA antibodies in the systemic circulation. No specific IgA in the fecal extract	Chew *et al.*, 2003

PLGA: poly(D,L-lactide-co-glycolide); SbutPVA-PLGA: sulfobutylated poly(vinylalcohol)-graft-poly(lactide-co-glycolide) PLA: poly (D,L-lactic acid); BSA : bovine serum albumin; NP: nanoparticles ; HE : Hot saline extract of *Salmonella* enteritidis.

4.2.3. Nanoparticles for Virus Encapsulation

Another interesting approach for oral vaccination with nanoparticles is the possibility to directly coat the microorganism with a polymer in order to protect it against inactivation and destruction during their residence in the stomach. In this context, Nechaeva [2002] proposed the encapsulation of live measles virus with pH-dependent polymers. The encapsulated form of live measles was constructed of particles with a size ranging from 100 to 1000 nm. Oral immunization of guinea pigs with particles of polyacrylic acid copolymers, polyacrylic acid-PVP, sodium alginate-chitosan or sodium alginate-spermidine demonstrated a significant increase in the antibody titers. These titers of protective antibodies were preserved in the circulation over 12 months after the immunization [Nechaeva, 2002].

4.2.4. Nanoparticles as Adjuvants for Oral Allergen-immunotherapy

Allergy is defined as "a hypersensitivity reaction mediated by immunological mechanisms". In the majority of cases the antibody typically responsible for an allergic reaction belongs to the IgE isotype and individuals may be referred to as suffering from an IgE-mediated allergic disease [Hamid and Minshall, 1996]. Nowadays, allergen-specifc immunotherapy (SIT) represents the only curative approach for allergies [Bernstein *et al.*, 2004]. Successful SIT appears to be associated with the counter-balance of the overwhelming allergic Th2 immune response due to (i) the upregulation of a Th1 immune response (immune deviation), (ii) the induction of a state of unresponsiveness (anergy) in peripheral T lymphocytes characterized by suppressed proliferative and cytokine responses against allergens, and (iii) the induction of regulatory T cells [Secrist *et al.*, 1993]. The modified T cell response subsequently affects the humoral allergen-specific immune response, i.e., the reduction of IgE and the induction of IgG4 antibodies with the capacity to block IgE-binding to the allergen [Francis *et al.*, 2003; Nouri-Aria *et al.*, 2004].

One approach that might be useful for immune therapy in allergic diseases could employ the induction of germinal centers in the gastrointestinal tract [Fagarasan *et al.*, 2002] in order to induce Th1 responses and the production of IgE blocking antibodies [Sprent and Tough, 2001; Weiner, 2001].

The "traditional" administration of allergens loaded in nanoparticles by the oral route has shown a very low efficacy to counterbalance the Th2 dominated immune response. This fact has been atributed to the absence of purified allergens and to the low specificity of conventional nanoparticles to reach the GALT. On the contrary, different studies have demonstrated that the oral delivery of plasmid DNA in nanoparticles can modify the immune system in mice and protect against food allergen-induced hypersensitivity [Roy *et al.*, 1999; Chew *et al.*, 2003]. This new oral allergen-gene immunization with chitosan-DNA nanoparticles has been proved effective in modulating anaphylactic responses, and indicated its prophylactic utility in treating peanut (food) [Roy *et al.*, 1999] and house dust mite (air) [Chew *et al.*, 2003] allergies.

In the first study, mice receiving nanoparticles containing a dominant peanut allergen gene (pCMVArah2) produced elevated secretory IgA and serum IgG2a titers. Compared with non-immunized mice or mice treated with "naked" DNA, the animals immunized with chitosan nanoparticles showed a substantial reduction in allergen-induced anaphylaxis associated with reduced levels of IgE, plasma histamine and vascular leakage [Roy *et al.*, 1999]. Similar results were obtained with plasmid DNA encoding genes of dominant allergens of dust mite [Chew *et al.*, 2003]. These DNA-nanoparticles, when delivered orally, could produce specific IgG2a and IgA antibodies in the systemic circulation of animals, whereas oral fed with naked DNA or intramuscular immunization alone could not. In any case, it is interesting to note that DNA protection by chitosan nanoparticles appeared to be crucial. Moreover, chitosan nanoparticles might also facilitate bioadhesion and DNA uptake by the host cells, leading to enhanced transfection efficiency.

4.3. Combination of Nanoparticles and Targeting Ligands

Unfortunately, oral immunization using conventional antigen-loaded nanoparticles is in general, not able to elicit a sufficient systemic immune response compared to parenteral routes, and consequently, high and multiple oral doses are required. This fact has been related with a low capacity of conventional antigen-loaded nanoparticles to target lymphoid tissues within the gut after oral administration.

The distribution of conventional nanoparticles within the gut is controlled by their physico-chemical properties and they do not offer the possibility to target or interact with specific regions such as M-cells of Peyer's patches [Ponchel and Irache, 1998]. For instance, for biodegradable PLGA nanoparticles, less than 0.01% of the given dose was found deposited in intestinal PPs of either rabbis or rats [McClean *et al.*, 1998]. In fact, conventional nanoparticles arc rapidly eliminated from the intestinal mucosa due to the continuous mucus turnover and intestinal peristaltism. In order to increase the residence of nanoparticles within the gut and to improve their capability to reach specific intestinal areas (including PPs and immunological cells), different strategies have been developed, either by modifying the surface properties of nanoparticles or by coupling a targeting molecule at their surface.

4.3.1. Modification of Nanoparticle Surface Properties

Modification of nanoparticle surface properties can be achieved either by coating nanoparticle surface with hydrophilic stabilizing agents, bioadhesive polymers or surfactants or by incorporating biodegradable copolymers containing hydrophilic residues in their structure. These modifications mainly change nanoparticle's zeta potential and their hydrophobicity, thus influencing the stability of the resulting nanoparticles and their ability to reach specific areas and develop adhesive interactions within the gut. One of the most popular techniques for modification of the nanoparticle surface is pegylation.

PEG has been employed as nanoparticle coating in drug delivery applications for its stabilizing properties. PEG chains form a steric barrier at the nanoparticle surface preventing protein binding, complement activation, and preferential uptake by the cells of the monocyte-macrophage system [Otsuka *et al.*, 2003]. Due to these interesting properties numerous PEGylated formulations have been developed [Takeuchi *et al.*, 2005; Vila *et al.*, 2002]. More recently, Yoncheva *et al.* [2005] developed pegylated Gantrez nanoparticles with a high ability to avoid adhesive interactions within the stomach and to concentrate them in the small intestine mucosa of animals. When these pegylated nanoparticles loaded with ovalbumin were administered by the oral route to Balb/c mice, discrete levels of specific antibodies were measured; although, an intense, sustained and prolonged induction of IL-10 in plasma was measured [Yoncheva *et al.*, 2005]. IL-10 has great impact on immunoregulation. It is a potent suppressor of several effector functions of macrophages, T cells and natural killer cells [Adorini, 2003]. Interestingly, its regulatory potential may be of interest in the treatment of autoimmune disorders [Adorini, 2003] and allergy [Blaser and Akdis, 2004].

4.3.2. Targeted Nanoparticles

Another strategy consists in grafting a ligand at nanoparticle surface to specifically target receptors expressed on the surface of enterocyte or M cells. Different types of targeting molecules have been tested including lectins [Irache *et al.*, 1994; Bies *et al.*, 2004], vitamin B_{12} derivatives [Russell-Jones *et al.*, 1999] and microorganism-derived adhesive factors [Salman *et al.*, 2005]. The most studied one has been the lectin family.

Lectins, which have been described as “second generation mucoadhesives” [Kompella and Lee, 1992] and have been proposed as tools for enhanced drug delivery to the gastrointestinal tract [Naisbett and Woodley, 1994], are natural proteins or glycoproteins that bind reversibly and specifically to sugars, and thus agglutinate cells and polysaccharides [Goldstein *et al.*, 1980]. In principle, their association to polymeric nanoparticles significantly increases their transport across the intestinal mucosa by efficiently increasing interactions with mucus and/or the surface of epithelial cells and by promoting particle translocation. In order to target the GALT different lectins have been proposed, including *Ulex europaus* lectin I [Ezpeleta *et al.*, 1996], *Sambucus nigra* [Arbos *et al.*, 2002] wheat germ agglutinin [Weissenbock *et al.*, 2004] and *Arachis hypogaea* agglutinin (AAL) [Gupta *et al.*, 2006]. Particularly interesting is a recent work in which Roth-Walker and co-workers [2005] have demonstrated that the coating of PLGA nanoparticles, encapsulating birch pollen antigens, with *Aleuria aurantia* lectin may be suitable for oral immunotherapy of Th2-dominated allergies. In this study, this formulation induced in mice a significant production of IgG2a antibodies (Th1 response) with no detection of IgG1 antibodies, whereas the contrary effect (high Th2 response) was found for conventional nanoparticles. Then, AAL-functionalized particles allow to accumulate antigens at the desired mucosal site after administration and to achieve a specific Th1 antibody response.

Other interesting approach to increase the capability of nanoparticles to target the gut-associated lymphoid tissue, may be to reproduce closely the bacteria and virus behavior concerning their strategies to infect and colonize the gut mucosa. In fact, bacterial interaction and adhesion is a prerequisite for epithelia colonization [Cossart, 2006] and, for this purpose, microorganisms have been developed a number of different approaches including the use of adhesins (flagella, pili and fimbria) and glycoconjugates (i.e. mannose proteins).

In this context, *Salmonella spp.* have the natural ability to invade both non-phagocytic host cells and Peyer's patches [Sirard *et al.*, 1999]. The real process by which *Salmonella* is able to colonize the gut mucosa is still unknown; although, different bacterial apendages, such as flagella play an important role in the gut mucosa colonization and invasion. Recently, "Salmonella-like" nanoparticles have been obtained by the association of *Salmonella* Enteritidis flagellin (the main component of flagellar filament) to Gantrez nanoparticles [Salman *et al.*, 2005]. These carriers displayed an important tropism for the ileum and their distribution within the gut correlated well with the described colonisation profile for Salmonella enteritidis, including a broad concentration in Peyer's patches. Using ovalbumin as model antigen, "Salmonella-like" nanoparticles induced a strong and balanced secretion of both IgG2a (Th1) and IgG1 (Th2) specific antibodies in the serum of animals. In addition, these nanoparticles were able to induce a much more strong mucosal IgA response than control particles [data not published].

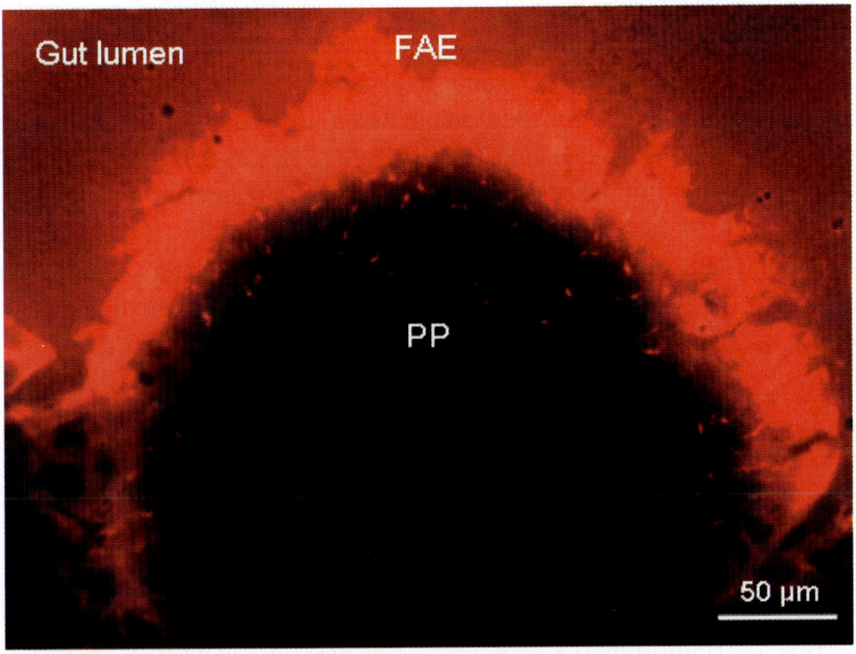

Fig. 3 Visualization by fluorescence microscopy of mannosylated nanoparticles (fluorescently labelled with rhodamine B isothyocianate) in the ileum of rats. FAE: follicle-associated epithelium; PP: Peyer's patch. Bar: 50 µm.

In addition to the well known microorganism's adhesive structures (i.e. flagellin and fimbriae), glycoconjugates enriched in mannose residues, expressed on the surface of microrganism's, have been found to promote their interaction with mucosal tissue of the gastrointestinal tract [Dalle *et al.*, 2003]. This adhesive mechanism may be related to the high binding affinity of mannose residues to the so-called mannose-binding lectins which are expressed on the lymphoid and non-lymphoid cells of the gut [Uemura *et al.*, 2002; Wagner *et al.*, 2003]. This strategy is found in many invasive microorganisms carrying mannose residues on their surfaces to favor their mucosal adhesion and gut colonization, including

Candida albicans, *Listeria monocytogenes*, HIV, some enterobacteria and bifidobacteria [reviewed in Kilpatrick, 2002].

Recently, mannose coating nanoparticles have been proposed for oral vaccination. Using ovalbumin as antigen model, these nanoparticles elicited a higher and more balanced specific antibody response (IgG1 and IgG2a) compared with conventional nanoparticles which elicited a typical Th2 response. In addition, mannose-coated nanoparticles were able to elicit a significant intestinal secretory IgA (s-IgA) for at least 6 weeks [data not published]. These immunogenic responses appear to be related with the ability of mannose-coated nanoparticles to specifically target the ileum of animals, including PP localized in this region (see Figure 3). In fact, the delivery of antigens to the distal small intestine region facilitates the induction of Th1 [Cronkhite and Michael, 2004].

5. Conclusions

In summary, oral immunization has a great potential, as outlined above. It can be exploited for the development of vaccines against mucosal acquired as well as non-mucosal acquired pathogens. Different studies have demonstrated that nanoparticles can represent an effective adjuvant system, inducing immune responses after oral administration. However, accumulated experimental evidence suggests that simple encapsulation of vaccines into nanoparticles is unlikely to result in the successful development of oral vaccines and improvements in the current technology are clearly needed. In any case, the use of nanoparticles is at an early research stage but appears promising for the future development of oral vaccines. We need to know more on the physiology of the mucosa and on the mechanisms and receptors that may favor an efficient targeting of vaccines to the mucosal immune system.

References

Adorini, L. (2003). Cytokine-based immunointervention in the treatment of autoimmune diseases. *Clin. Exp. Immunol.*, 132, pp. 185-192.

Arbos, P., Wirh M., Arangoa M. A., Gabor F. and Irache J. M. (2002). Gantrez® AN as a new polymer for the preparation of ligand-nanoparticle conjugates. *J. Control. Rel.*, 83, pp. 321-330.

Arbós, P., Campanero, M. A., Arangoa, M. A., Renedo, M. J. and Irache, J. M. (2003). Influence of the surface characteristics of PVM/MA nanoparticles on their bioadhesive properties. *J. Control. Rel.*, 89, pp. 19-30.

Ardavin, C. (2003). Origin, precursors and differentiation of mouse dendritic cells. *Nat. Rev. Immunol.*, 3, pp. 582-590.

Banchereau, J. and Steinman, R. M. (1998). Dendritic cells and the control of immunity. *Nature*, 392, pp. 245-252.

Bernstein, D. I., Wanner, M., Borish, L. and Liss, G. M. (2004). Twelve-year survey of fatal reactions to allergen injections and skin testing: 1990-2001. *J. Allergy Clin. Immunol.*, 113, pp. 1129-1136.

Bies, C., Lehr, C. M. and Woodley, J. F. (2004). Lectin-mediated drug targeting: history and applications. *Adv. Drug Deliv. Rev.*, 56, pp. 425-435.

Bivas-Benita, M., Laloup, M., Versteyhe, S., Dewit, J., De, B. J., Jongert, E. and Borchard, G. (2003). Generation of Toxoplasma gondii GRA1 protein and DNA vaccine loaded chitosan particles: preparation, characterization, and preliminary in vivo studies. *Int. J. Pharm.*, 266, pp. 17–27.

Blaser, K. and Akdis, A. (2004). Interleukin-10, T regulatory cells and specific allergy treatment. *Clin. Exp. Allergy*, 34, 328-331.

Brayden, D. J. (2001). Oral vaccination in man using antigens in particles: current status. *Eur. J. Pharm. Sci.*, 14, pp. 183-189.

Brayden, D. J. and Baird, A. W. (2004). Apical membrane receptors on intestinal M cells: potential targets for vaccine delivery. *Adv. Drug Deliv. Rev.*, 56, pp. 721–726.

Carcaboso, A. M., Hernandez, R. M., Igartua, M., Gascon, A. R., Rosas, J. E., Patarroy, M. E. and Pedraz, J. L. (2003). Immune response after oral administration of the encapsulated malaria synthetic peptide SPf66. *Int. J. Pharm.*, 260, pp. 273-282.

Chew, J. L., Wolfowicz, C. B., Mao, H. Q., Leong, K. W. and Chua, K. Y. (2003). Chitosan nanoparticles containing plasmid DNA encoding house dust mite allergen, Der p 1 for oral vaccination in mice. *Vaccine*, 21, pp. 2720-2729.

Clements, C. J. and Griffiths, E. (2002). The global impact of vaccines containing aluminium adjuvants. *Vaccine*, 20, pp. S24-S33.

Coffman, R. L., Varkila, K., Scott, P. and Chatelain, R. (1991). Role of cytokines in the differentiation of CD4+ T-cell subsets in vivo. *Immunol. Rev.*, 123, pp. 189-207.

Conway M. A., Madrigal-Estebas, L., McClean, S., Brayden, D. J. and Mills, K. H. (2001). Protection against Bordetella pertussis infection following parenteral or oral immunization with antigens entrapped in biodegradable particles: effect of formulation and route of immunization on induction of Th1 and Th2 cells. *Vaccine*, 19, pp.1940–1950.

Cossart, P. (2004). Bacterial invasion: a new strategy to dominate cytoskeleton plasticity. *Dev. Cell*, 6, pp. 314-315.

Cronkhite, R. I. and Michael, J. G. (2004) Sub-compartmentalization of the gastrointestinal (GI) immune system determined with microbeads that differ in release properties, *Vaccine*, 22, pp. 2106-2115

Dalle, F., Jouault, T., Trinel, P. A., Esnault, J., Mallet, J. M., d'Athis, P., Poulain D. and Bonnin, A. (2003). Beta-1,2- and alpha-1,2-linked oligomannosides mediate adherence of Candida albicans blastospores to human enterocytes in vitro. Infect. *Immun.*, 71, pp. 7061-7068.

De Magistris, M. T. (2006). Mucosal delivery of vaccine antigens and its advantages in pediatrics, *Adv. Drug Deliv. Rev.*, 58, pp. 52-67.

Degen, W. G., Jansen, T. and Schijns, V. E. (2003). Vaccine adjuvant technology: from mechanistic concepts to practical applications. *Expert. Rev. Vaccines*, 2, pp. 327-335.

Desai, M. P., Labhasetwar, V., Amidon G. L. and Levy, R. J. (1996). Gastrointestinal uptake of biodegradable microparticles: effect of particle size. *Pharm. Res.*, 13, pp. 1838–1845.

Durrer, P., Gluck, U., Spyr, C., Lang, A. B., Zurbriggen, R., Herzog, C. and Gluck, R., (2003). Mucosal antibody response induced with a nasal virosome-based influenza vaccine. *Vaccine*, 21, pp. 4328– 4334.

Eldridge, J. H., Meulbroek, J. A., Staas, J. K., Tice, T. R. and Gilley, R. M (1989). Vaccine-containing biodegradable microspheres specifically enter the gut-associated lymphoid tissue following oral administration and induce a disseminated mucosal immune response. *Adv. Exp. Med. Biol.*, 251, pp. 191–202.

Ensoli, B., Barillari, G., Salahuddin, S. Z., Gallo, R. C. and Wong-Staal, F. (1999). Tat protein of HIV-1 stimulates growth of cells derived from Kaposi's sarcoma lesions of AIDS patients. *Nature*, 345, pp. 84-86.

Espuelas, S., Irache, J. M. and Gamazo, C. (2005). Synthetic particulate antigen delivery systems for vaccination. *Inmunología*, 24, pp. 207-223.

Ezpeleta, I., Irache, J. M., Stainmesse, S., Chabenat, C., Gueguen, J. and Orecchioni, A. M. (1996). Preparation of lectin-vicilin nanoparticle conjugates using the carbodiimide coupling technique. *Int. J. Pharm.*, 142, pp. 227-233.

Fagarasan, S., Muramatsu, M., Suzuki, K., Nagaoka, H., Hiai, H. and Honjo, T. (2002). Critical roles of activation-induced cytidine deaminase in the homoeostasis of gut flora. *Science*, 298, pp. 1424–1427.

Fernandez, M. I., Pedron, T., Tournebize, R., Olivo-Marin, J. C., Sansonetti, P. J. and Phalipon, A. (2003). Anti-inflammatory role for intracellular dimeric immunoglobulin a by neutralization of lipopolysaccharide in epithelial cells. *Immunity*, 18, pp. 739– 749.

Francis, J. N., Till, S. J. and Durham, S. R. (2003). Induction of IL-10+CD4+CD25+ T cells by grass pollen immunotherapy. *J. Allergy Clin. Immunol.*, 111, pp. 1255-1261.

Gamazo, C., Ochoa-Repáraz, J. and Irache, J. M. (2004). Efficacy of a vaccine against *Salmonella enteritidis* based on Gantrez nanoparticles as adjuvant and delivery system. *Clin. Microbiol. Infect.*, 10, pp. 332-333.

Goldstein, I. J., Hughes, R. C., Monsigny, M., Osawa, T. and Sharon, N. (1980). What should be called lectin? *Nature*, 285, pp. 66.

Gupta, P. N., Mahor, S., Rawat, A., Khatri, K., Goyal, A. and Vyas, S. P. (2006). Lectin anchored stabilized biodegradable nanoparticles for oral immunization 1. Development and in vitro evaluation. *Int. J. Pharm.*, 318, pp.163-173.

Gutierro I., Hernandez, R. M., Igartua, M., Gascon, A. R. and Pedraz, J. L. (2002 a). Size dependent immune response after subcutaneous, oral and intranasal administration of BSA loaded nanospheres. *Vaccine*, 21, pp. 67–77.

Gutierro, I., Hernandez, R. M., Igartua, M., Gascon, A. R. and Pedraz, J. L. (2002 b). Influence of dose and immunization route on the serum Ig G antibody response to BSA loaded PLGA microspheres. *Vaccine*, 20, pp. 2181–2190.

Hamada, H., Hirió, T., Nishiyama, Y., Takahashi H., Yohei Masunaga Y., Hachimura, S., Kaminogawa, S., Takahashi-Iwanaga, H., Iwanaga, T., Kiyono, H., Yamamoto, H. and Ishikawa, H. (2002). Identification of multiple isolated lymphoid follicles on the antimesenteric wall of the mouse small intestine. *J. Immunol.*, 168, pp. 57–64.

Hamid, Q. A. and Minshall, E. (1996). In situ detection of cytokines in allergic inflammation. *Adv. Exp. Med. Biol.*, 409, pp. 327-335.

Herrmann, J. E., Chen, S. C., Jones, D. H., Tinsley-Bown, A., Fynan, E. F., Greenberg, H. B. and Farrar, G. H. (1999). Immune responses and protection obtained by oral immunization with rotavirus VP4 and VP7 DNA vaccines encapsulated in microparticles. *Virology*, 259, pp.148–153.

Huckriede A., Bungener, L., Daemen, T. and Wilschut, J. (2003). Influenza virosomes in vaccine development. *Methods Enzymol.*, 373, pp. 74-91.

Iijima, H., Takahashi, I. and Kiyono, H. (2001). Mucosal immune network in the gut for the control of infectious diseases. *Rev. Med. Virol.*, 11, pp. 117–133.

Irache, J. M., Durrer, C., Duchene, D. and Ponchel, G. (1994). Preparation and characterization of lectin-latex conjugates for specific adhesion. *Biomaterials*, 11, pp. 899-904.

Iwasaki, A. and Kelsall, B. L. (2000). Localization of distinct Peyer's patch dendritic cell subsets and their recruitment by chemokines macrophage inflammatory protein (MIP)-3α, MIP-3β, and secondary lymphoid organ chemokine. *J. Exp. Med.*, 191, pp. 1381–1393.

Iwasaki, A. and Kelsall, B. L. (2001). Unique function of CD11b+, CD8a+, and double-negative Peyer's patch dendritic cells. *J. Immunol.*, 166, pp. 4884–4890.

Jain, S., Sharma, R. K. and Vyas, S. P. (2006). Chitosan nanoparticles encapsulated vesicular systems for oral immunization: preparation, in-vitro and in-vivo characterization. *J. Pharm. Pharmacol.*, 58, pp. 303-310.

Jepson, M. A., Simmons, N. L., Hirst, G. L. and Hirst, B. H. (1993). Identification of M cells and their distribution in rabbit intestinal Peyer's patches and appendix. *Cell Tissue Res.*, 273, pp. 127-136.

Jones, D. H., Corris, S., McDonald, S., Clegg, J. C. and Farrar, G. H. (1997). Poly(DL-lactide-co-glycolide)-encapsulated plasmid DNA elicits svstemic and mucosal antibody responses to encoded protein after oral administration. *Vaccine*, 15, pp. 814-817.

Katz, D. E., DeLorimier, A. J., Wolf, M. K., Hall, E. R., Cassels, F. J., van Hamont, J. E., Newcomer, R. L., Davachi, M. A., Taylor, D. N. and McQueen, C. E. (2003). Oral immunization of adult volunteers with microencapsulated enterotoxigenic *Escherichia coli* (ETEC) CS6 antigen. *Vaccine*, 21, pp. 341-346.

Kelsall, B. L. and Strober, W. (1996). Distinct populations of dendritic cells are present in the subepithelial dome and T cell regions of the murine Peyer's patch. *J. Exp. Med.*, 183, pp. 237–247.

Kersten, G., Drane, D., Pearse, M., Jiskoot, W. and Coulter A. (2004). "Liposomes and ISCOMS" in *Novel Vaccination Strategies*, eds. Kaufmann, S., (Wiley-VCH: Weinheim), pp. 173-196.

Kilpatrick, D. C. (2002). Mannan-binding lectin: clinical significance and applications. *Biochim. Biophys. Acta*, 1572, pp. 401–413.

Kim, S. Y., Doh, H. J., Jang, M. H., Ha, Y. J., Chung, S. I. and Park, H. J. (1999). Oral immunization with *Helicobacter pylori*-loaded poly(D,L-lactide-co-glycolide) nanoparticles. *Helicobacter*, 4, pp. 33-39.

Kompella, U. B. and Lee, V. (1992). Means to enhance penetration. 4. Delivery systems for the enhancement of peptide and protein drugs: design considerations. *Adv. Drug Del. Rev.*, 8, pp. 115-162.

Krahenbuhl, J. P. and Neutra, M. R. (2000). Epithelial M cells: differentiation and function. *Annu. Rev. Cell Dev. Biol.*, 16, pp. 301-332.

Kunkel, E. J. and Butcher, E. C. (2003). Plasma-cell homing. *Nat. Rev. Immunol.*, 3, pp. 822-829.

Le Buanec H., Vetu, C., Lachgar, A., Benoit, M. A., Gillard, J., Paturance, S., Aucoututier, J., Gane, V., Zaguri, D. and Bizzini, B. (2001). Induction in mice of anti-Tat mucosal immunity by the intranasal and oral routes. *Biomed. Pharmacother.*, 55, pp. 316-320.

Levine, M. M., Woodrow, G. C., Kaper, J. B. and Cobon, G. S. (1997). *New generation vaccines*, 2nd Edn. (Marcel Dekker Inc., New York, USA).

Levine, M. M., (2003). Can needle-free administration of vaccines become the norm in global immunization? *Nat. Med.*, 9, pp. 99-103.

Lillard, J. W. Jr., Boyaka, P. N., Chertov, O., Oppenheim, J. J. and McGhee, J. R. (1999). Mechanisms for induction of acquired host immunity by neutrophil peptide defensins, *Proc. Natl. Acad. Sci.*, 96, pp. 651-656.

Lindblad, E. B. (2004). Aluminium compounds for use in vaccines. *Immunol. Cell. Biol.*, 82, pp. 497-505.

London, S. D., Rubin, D. H. and Cebra, J. J. (1987). Gut mucosal immunization with reovirus serotype 1/L stimulates virus-specific cytotoxic T cell precursors as well as IgA memory cells in Peyer's patches. *J. Exp. Med.*, 165, pp. 830-847.

Lutsiak, M. E., Robinson, D. R., Coester, C., Kwon, G. S. and Samuel, J. (2002). Analysis of poly(d,l-lactic-co-glycolic acid) nanosphere uptake by human dendritic cells and macrophages in vitro. *Pharm. Res.*, 19, pp. 1480-1487.

Maloy, K. J., Donachie, A. M., O'Hagan, D. T. and Mowat, A. M. (1994). Induction of mucosal and systemic immune responses by immunization with ovalbumin entrapped in poly(lactide-co-glycolide) microparticles. *Immunology*, 81, pp. 661-667.

Mann, J. F., Scales, H. E., Shakir, E., Alexander, J., Carter, K. C., Mullen, A. B. and Ferro, V. A. (2006). Oral delivery of tetanus toxoid using vesicles containing bile salts (bilosomes) induces significant systemic and mucosal immunity. *Methods*, 38, pp. 90-95.

Marx, P. A., Compans, R. W. and Gettie, A. (1993). Protection against vaginal SIV transmition with microencapsulated vaccine. *Science*, 28, pp. 1323–1327.

Mazumdar T., Anam, K. and Ali, N. (2005). Influence of phospholipid composition on the adjuvanticity and protective efficacy of liposome-encapsulated *Leishmania donovani* antigens. *J. Parasitol.*, 91, pp. 269-274.

McClean, S., Prosser, E., Meehan, E., O'Malley, D., Clarke, N., Ramtoola, Z. and Brayden, D. J. (1998). Binding and uptake of biodegradable poly-dl-lactide micro- and nanoparticles in intestinal epithelia. *Eur. J. Pharm. Sci.*, 6, pp. 153-163.

McGhee, J. R., Mestecky, J., Dertzbaugh, M. T., Eldridge, J. H., Hirasawa, M. and Kiyono, H. (1992). The mucosal immune system: from fundamental concepts to vaccine development. *Vaccine*, 10, pp. 75-88.

Miyoshi, J. and Takai, Y. (2005). Molecular perspective on tight-junction assembly and epithelial polarity. *Adv. Drug Deliv. Rev.*, 57, pp. 815-855.

Morein, B., Lövgren, K., Rönnberg, B., Sjölander, A. and Villacres-Eriksson, M. (1995). Immunostimulating complexes: Clinical potential in vaccine development. *Clin. Immunother.*, 3, pp. 461-475.

Morein, B., Sundquist, B., Hoglund, S., Dalsgaard K. and Osterhaus, A. (1984). Iscom, a novel structure for antigenic presentation of membrane proteins from enveloped viruses. *Nature*, 308, pp. 457-460.

Morein, B., Villacres-Eriksson, M., Åkerblom, L., Rönnberg, B., Lövgren, K. and Sjölander, A. (1994). Mechanisms behind the immune response induced by immunostimulating complexes. *Aids Res. Hum. Retrovir.*, 10, pp. S109-S114.

Moser, C., Metcalfe, I. C. and Viret, J. F. (2003). Virosomal adjuvanted antigen delivery systems. *Expert Rev. Vaccines*, 2, pp. 189-196.

Murillo, M., Goñi, M. M., Irache, J. M., Arangoa, M. A., Blasco, J. M. and Gamazo, C. (2002). Modulation of the cellular immune response after oral or subcutaneous immunization with microparticles containing Brucella *ovis* antigens. *J. Control. Rel.*, 85, pp. 237-246.

Nagler-Anderson, C. (2001). Man the barrier! Strategic defences in the intestinal mucosa. *Nat. Rev. Immunol.*, 1, pp. 59-67.

Naisbett, B. and Woodley, J. (1994). The potential of tomato lectin for oral drug delivery. 2. Mechanism of uptake in vitro. *Int. J. Pharm.*, 110, pp. 127-136.

Nechaeva, E. (2002). Development of oral microencapsulated forms for delivering viral vaccines. *Exp. Rev. Vaccines*, 1, pp. 385-397.

Neutra, M. R., Frey, A. and Kraehenbuhl, J. P. (1996). Epithelial M cells: gateways for mucosal infection and immunization. *Cell*, 86, pp. 345-348.

Newman, K. D., Elamanchili, P., Kwon, G. S. and Samuel, J. (2002). Uptake of poly(d,l-lactic-co-glycolic acid) microspheres by antigen-presenting cells in vivo. *J. Biomed. Mater. Res.*, 60, pp. 480-486.

Nouri-Aria, K. T., Wachholz, P. A., Francis, J. N., Jacobson, M. R., Walker, S. M., Wilcock, L. K., Staple, S. Q., Aalberse, R. C., Till, S. J. and Durham, S. R. (2004). Grass pollen immunotherapy induces mucosal and peripheral IL-10 responses and blocking IgG activity. *J. Immunol.*, 172, pp. 3252-3259.

O'Hagan, D. T., McGee, J. P., Lindblad, M. and Holmgren, J. (1995). Cholera toxin B subunit (CTB) entrapped in microparticles shows comparable immunogenicity to CTB mixed with whole cholera toxin following oral immunization. *Int. J. Pharm.*, 119, pp. 251-255.

O'Hagan, D. T. and Valiante, N. M. (2003). Recent advances in the discovery and delivery of vaccine adjuvants. *Nat. Rev. Drug Discov.*, 2, pp. 727-735.

Okada, H. and Toguchi, H. (1995). Biodegradable microspheres in drug delivery. *Crit. Rev. Ther. Drug Carr. Syst.*, 12, pp. 1-99.

Olbrich, C., Muller, R. H., Tabatt, K., Kayser, O., Schulze, C. and Schade, R. (2002). Stable biocompatible adjuvants--a new type of adjuvant based on solid lipid nanoparticles: a study on cytotoxicity, compatibility and efficacy in chicken. *Altern. Lab. Anim.*, 30, pp. 443-458.

Otsuka, H., Nagasaki, Y. and Kataoka, K. (2003). PEGylated nanoparticles for biological and pharmaceutical applications. *Adv. Drug Deliv. Rev.*, 55, pp. 403-419.

Otterlei, M., Varum, K. M., Ryan, L. and Espevik, T. (1994). Characterization of binding and TNF-alpha-inducing ability of chitosans on monocytes: the involvement of CD14. *Vaccine*, 12, pp. 825-832.

Owen, R. L. and Jones, A. L. (1974). Epithelial cell specialization within human Peyer's patches; and ultrastructural study of internal lymphoid follicles. *Gastroenterology*, 66, pp. 189-203.

Papadopoulou, G., Karagouni, E. and Dotsika, E. (1998). ISCOMs vaccine against experimental leishmaniasis. *Vaccine*, 16, pp. 885-892.

Pearse, J. and Drane, D. (2005). Iscomatrix® adjuvant for antigen delivery. *Adv. Drug Del. Rev.*, 57, pp. 465-474.

Peppoloni, S., Ruggiero, P., Contorni, M., Morandi, M., Pizza, M., Rappuoli, R., Podda, A. and Del Giudice, G. (2003). Mutants of the *Escherichia coli* heat-labile enterotoxin as safe and strong adjuvants for intranasal delivery of vaccines. *Expert Rev. Vaccines*, 2, pp. 285-293.

Perez, O., Bracho, G., Lastre, M., Mora, N., del Campo, J., Gil, D., Zayas, C., Acevedo, R., Gonzalez, D., Lopez, J. A., Taboada, C., Turtle, C. and Solis, R. L. (2004). Novel adjuvant based on a proteoliposome-derived cochleate structure containing native lipopolysaccharide as a pathogen-associated molecular pattern. *Immunol. Cell. Biol.*, 82, pp. 603-610.

Petrovsky, N. and Aguilar, J. C. (2004). Vaccine adjuvants: current state and future trends. *Immunol. Cell. Biol.*, 82, pp. 488-496.

Ponchel, G. and Irache, J. M. (1998). Specific and non-specific bioadhesive particulates for oral delivery. *Adv. Drug Del. Rev.*, 34, pp. 191-219.

Porporatto, C., Bianco, I. D. and Correa, S. G. (2005). Local and systemic activity of the polysaccharide chitosan at lymphoid tissues after oral administration. *J. Leukoc. Biol.*, 78, pp. 62-69.

Poulain-Godefroy, O., Mielcarek, N., Ivanoff, N., Remoué, F., Schacht, A. M., Phillips, N., Locht, C., Capron, A. and Riveau, G. (1998). Bordetella pertussis filamentous hemagglutinin enhances the immunogenicity of liposome-delivered antigen administered intranasally. *Infect. Immun.*, 66, pp. 1764-1767.

Ramon, G. (1925). Sur l'augmentation anormale de l'antitoxine chez les chevaux producteurs de serum antidiphterique. *Bull. Soc. Centr. Med. Vet.*, 101, pp. 227-234.

Ramon, G. (1926) Procédés pour accroître la production des antitoxins. *Ann. Inst. Pasteur*, 40, pp. 1-10.

Rescigno, M., Urbano, M., Valzasina, B., Francolini, M., Rotta, G., Bonasio, R., Granucci, F., Kraehenbuhl, J. P. and Ricciardi-Castagnoli, P. (2001). Dendritic cells express tight junction proteins and penetrate gut epithelial monolayers to sample bacteria. *Nat. Immunol.*, 2, 361-367.

Rojas, R. and Apodaca, G. (2002). Immunoglobulin transport across polarized epithelial cells. *Nat. Rev. Mol. Cell Biol.*, 3, pp. 944-955.

Rosas, J. E., Hernández, R. M., Gascón, A. R., Igartua, M., Guzmán, F., Patarroyo M. E. and Pedraz, J. L. (2001). Biodegradable PLGA microspheres as a delivery system for malaria synthetic peptide SPf66a. *Vaccine*, 19, pp. 4445-4451.

Roth-Walter, F., Scholl, I., Untersmayr, E., Ellinger, A., Boltz-Nitulescu, G., Scheiner, O., Gabor, F. and Jensen-Jarolim, E. (2005). Mucosal targeting of allergen loaded microspheres by Aleuria aurantia lectin. *Vaccine*, 23, pp. 2703-2710.

Roy K., Mao, H. Q., Huang, S. K. and Leong, K. W. (1999). Oral gene delivery with chitosan-DNA nanoparticles generates immunologic protection in a murine model of peanut allergy. *Nat. Med.*, 5, pp. 387-391.

Russell-Jones, G. J., Arthur L. and Walker, H. (1999). Vitamin B12-mediated transport of nanoparticles across Caco-2 cells. *Int. J. Pharm.*, 179, pp. 247-255.

Salman, H. H., Gamazo, C., Campanero M. A. and Irache, J. M. (2005). Salmonella-like bioadhesive nanoparticles. *J. Control. Rel.*, 106, pp. 1-13.

Schijns, V. E. (2003). Mechanisms of vaccine adjuvant activity: initiation and regulation of immune responses by vaccine adjuvants. *Vaccine*, 21, pp. 829-831.

Secrist, H., Chelen, C. J., Wen, Y., Marshall, J. D. and Umetsu, D. T. (1993). Allergen immunotherapy decreases interleukin 4 production in CD4+ T cells from allergic individuals. *J. Exp. Med.*, 178, pp. 2123-2130.

Shakweh, M., Ponchel, G. and Fattal, E. (2004). Particle uptake by Peyer's patches: a pathway for drug and vaccine delivery. *Expert. Opin.. Drug Deliv.*, 1, pp. 141-163.

Sheikh, N. A., al-Shamisi, M. and Morrow, W. J. (2000). Delivery systems for molecular vaccination. *Curr. Opin. Mol. Ther.*, 2, pp. 37-54.

Singh, M. and O'Hagan, D. T. (1999). Advances in vaccine adjuvants. *Nature Biotechnol.*, 17, pp. 1075-1081.

Singh, M. and O'Hagan, D. T. (2003). Recent advances in veterinary vaccine adjuvants. *Int. J. Parasitol.*, 33, pp. 469-478.

Sirard, J. C., Niedergang, F. and Kraehenbuhl, J. P. (1999). Live attenuated Salmonella: a paradigm of mucosal vaccines, *Immunol. Rev.*, 171, pp. 5-26.

Smith, C. M., Wilson, N. S., Waithman, J., Villadangos, J. A., Carbone, F. R., Heath, W. R. and Belz, G. T. (2004). Cognate CD4(+) T cell licensing of dendritic cells in CD8(+) T cell immunity. *Nat. Immunol.*, 5, pp. 1143-1148.

Sprent, J. and Tough, D. F. (2001). T cell death and memory. *Science*, 293, pp. 245-248.

Storni, T., Kundig, T. M. Senti, G. and Johansen, P. (2005). Immunity in response to particulate antigen-delivery systems. *Adv. Drug Deliv. Rev.*, 57, pp. 333-355.

Tabata, Y., Inoue, Y. and Ikada, Y. (1996). Size effect on systemic and mucosal immune responses induced by oral administration of biodegradable microspheres. *Vaccine*, 14, pp. 1677-1685.

Takeuchi, H., Thongborisute, J., Matsui, Y., Sugihara, H., Yamamoto, H. and Kawashima, Y. (2005). Novel mucoadhesion tests for polymers and polymer-coated particles to design optimal mucoadhesive drug delivery systems. *Adv. Drug Deliv. Rev.*, 57, pp. 1583-1594.

Uemura, K., Ska, M., Nakagawa, T., Kawasaki, N., Thiel, S., Jensenius, J. C. and Kawasaki, T. (2002). L-MBP is expressed in epithelial cells of mouse small intestine. *J. Immunol.*, 169, pp. 6945-6950.

Ulmer, J. B. (2004). Enhancement of vaccine potency through improved delivery. *Expert Opin. Biol. Ther.*, 4, pp. 1045-1051.

Vila, A., Sanchez, A., Tobio, M., Calvo, P. and Alonso, M. J. (2002). Design of biodegradable particles for protein delivery. *J. Control. Rel.*, 78, pp. 15-24.

Wagner, S., Lynch, N. J., Walter, W., Schwaeble W. J. and Loos, M. (2003). Differential expression of the murine mannose-binding lectins A and C in lymphoid and non lymphoid organs and tissues. *J. Immunol.*, 170, pp. 1462-1465.

Weiner, H. L. (2001). Induction and mechanism of action of transforming growth factor-beta-secreting Th3 regulatory cells. *Immunol. Rev.*, 182, pp. 207-214.

Weinstein, D. L., O'Neill, B. L., Hone, D. M. and Metcalf, E. S. (1998). Differential early interactions between Salmonella enterica serovar Typhi and two other pathogenic Salmonella serovars with intestinal epithelial cells. *Infect. Immun.*, 66, pp. 2310-2318.

Weissenbock, A., Wirth, M. and Gabor, F. (2004). WGA-grafted PLGA-nanospheres: preparation and association with Caco-2 single cells. *J. Control. Rel.*, 99, pp. 383-392.

Williams, I. R. (2004). Chemokine receptors and leukocyte trafficking in the mucosal immune system. *Immunol. Res.*, 29, pp. 283-292.

Yoncheva, K., Lizarraga, E. and Irache, J. M. (2005). Pegylated nanoparticles based on poly(methyl vinyl ether-co-maleic anhydride): preparation and evaluation of their bioadhesive propertie. *Eur. J. Pharm. Sci.*, 24, pp. 411-419.

Chapter 7

NANOPARTICLES: THERAPEUTIC APPROACHES FOR BACTERIAL DISEASES

Brice Moulari

Laboratory of Pharmaceutical Engineering,
Faculty of Medicine and Pharmacy, University of Franche-Comté,
Besançon (France)

1. Introduction

Antibiotics (Greek anti, "against"; bios, "life") are chemical compounds used to kill or inhibit the growth of bacteria. Antibiotics are one class of antimicrobials, a larger group which also includes anti-viral, anti-fungal, and anti-parasitic drugs. They are relatively harmless to the host, and therefore can be used to treat infections. The term originally described only those formulations derived from living organisms, in contrast to "chemotherapeutic agents", which are purely synthetic. Nowadays the term "antibiotic" is also applied to synthetic antimicrobials. Antibiotics are generally small molecules with a molecular weight less than 2000Da. They are not enzymes. Some antibiotics have been derived from mould, for example the penicillin class.

Conventional antibiotics are not effective in viral, fungal and other nonbacterial infections, and individual antibiotics vary widely in their effectiveness on various types of bacteria. The effectiveness of individual antibiotics varies with the location of the infection, the ability of the antibiotic to reach the site of infection, and the ability of the bacteria to resist or inactivate the antibiotic. However, use or misuse of antibiotics may result in the development of antibiotic resistance by the infecting organisms. With the development of bacteria resistance to antibiotics, there is considerable interest, on the one hand, for the development of other classes of antibiotics for the control of infection; and, on the other hand for increase the bioavailability of existing antibiotics.

In the case of increase the bioavailability of existing antibiotics, the solution consists to associate the antibiotic drug to a submicroscopic carrier thereby hiding and protecting the molecule from degradation and delivering it to inaccessible target in a controlled manner. Nanoparticles are carriers developed for these logistic targeting strategies and are colloidal in nature biodegradable and similar in behaviour to intracellular pathogens. Indeed, when administered by the intravenous route, polymeric particles localize preferentially in organs with high phagocytic activity and in circulating monocytes, ensuring their clearance [Poste, 1983; Grislain *et al.*, 1983]. The ability of circulating carriers to target these cells is highly dependent on tissue characteristics and on the carrier's properties.

2. Generality on Antibiotics

Many ancient cultures already used moulds and plants to treat infections. Modern research on antibiotics began with the discovery of Penicillin, derivative of the mold Penicillium notatum, in 1928 by Alexander Fleming. This discovery marked the beginning of the development of antibacterial compounds produced by living organisms. Indeed, antibiotic was originally used to refer only to substances extracted from a fungus or other microorganism, but has come to include also many synthetic and semi-synthetic drugs that have antibacterial effects.

The most common method classifies them according to their action against the infecting organism. Some antibiotics attack the cell wall; some disrupt the cell membrane; and the majority inhibit the synthesis of nucleic acids and proteins, the polymers that make up the bacterial cell. Another method classifies antibiotics according to which bacterial strains they affect: staphylococcus, streptococcus, or *Escherichia coli,* for example. Antibiotics are also classified on the basis of chemical structure, as penicillins, cephalosporins, aminoglycosides, tetracyclines, macrolides, or sulfonamides, among others.

Antibiotics may also be classed as bactericidal (killing bacteria) or bacteriostatic (stopping bacterial growth and multiplication). Bacteriostatic drugs are nonetheless effective because bacteria that are prevented from growing will die off after a time or be killed by the defense mechanisms of the host. The tetracyclines and the sulfonamides are among the bacteriostatic antiobiotics. Antibiotics that damage the cell membrane cause the cell's metabolites to leak out, thus killing the organism. Such compounds, including penicillins and cephalosporins, are therefore classed as bactericidal.

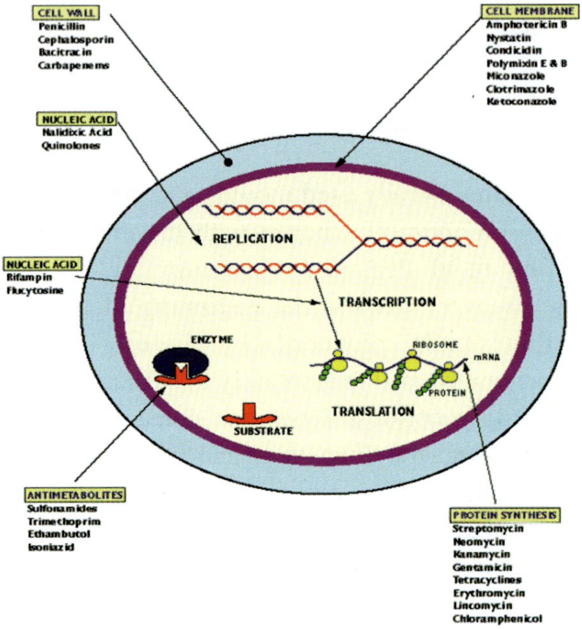

Fig. 1 Mechanisms of action of antibiotics.

The process of production usually involves screening of wide ranges of microorganisms, testing and modification. Production is carried out using fermentation. The naturally fermented product may be modified chemically to produce a semi synthetic antibiotic. After purification, the effect of the antibiotic on the normal function of host tissues and organs (its pharmacology) as well as its possible toxic actions (toxicology), must be tested on a large number of animals of several species.

These procedures, from the time the antibiotic is discovered in the laboratory until it undergoes clinical trial, usually extend over several years.

Being such an important medical compound the mechanisms by which antibiotics kill bacteria have been under scrutiny for decades, with such studies being instrumental in the design of new and improved compounds. There are three general modes of antibiotic activity:

(1) interference with the cell wall, (2) interference with nucleic acid synthesis, and (3) interference with protein synthesis (Figure 1).

Use or misuse of antibiotics may result in development of antibiotic resistance by the bacterial organisms. One of the main mechanisms of resistance is inactivation of the antibiotic. This is the usual resistance against penicillins and chloramphenicol, among others. Another form of resistance involves a mutation that changes the bacterial enzyme affected by the drug in such a way that the antibiotic can no longer inhibit it. This is the main mechanism of resistance to the compounds that inhibit protein synthesis, such as the tetracyclines.

The problem of resistance has been exacerbated by the use of antibiotics as prophylactics, intended to prevent infection before it occurs. Indiscriminate and inappropriate use of antibiotics for the treatment of the common cold and other common viral infections, against which they have no effect, removes antibiotic-sensitive bacteria and allows the development of antibiotic-resistant bacteria.

3. Antibiotics and Nanoencapsulation

Nanoparticles are receiving considerable attention for the delivery of antibiotic drugs. Indeed, resistance of bacteria to antibiotics has increased in recent years due to the development of resistant strains. Some antibiotic agents are extremely irritant and toxic and there is much interest in finding ways to formulate new types of safe and cost-effective antibacterial materials. Many studies have shown that antibacterial formulations in the nanoparticles form could be used as effective bactericidal materials [Lboutounne *et al.*, 2002; Qi *et al.*, 2004; Moulari *et al.*, 2005 and 2006; Li *et al.*, 2006; Beyth *et al.*, 2006; Nhung *et al.*, 2007; Turos *et al.*, 2007; Grace and Pandian, 2007].

In addition difficulty with classical antibiotic therapy is that many intracellular bacteria are quiescent or dormant. These bacteria are present in a reversible state and can persist for extended periods without division [Kaprelyants et al., 1993] under a viable but non-culturable state. So the drug targets are down regulated and the cellular permeability is altered dramatically, influencing their susceptibility to antibacterial agents

(Gilbert *et al.*, 1990). Indeed, microorganisms in infected tissues are protected by various biological structures around the infection foci.

Then despite the discovery of new antibiotics, the treatment of intracellular infections often completely fails to eradicate the pathogens. By loading antibiotics into nanoparticles, one can expect improved delivery to infected cells (Figure 2).

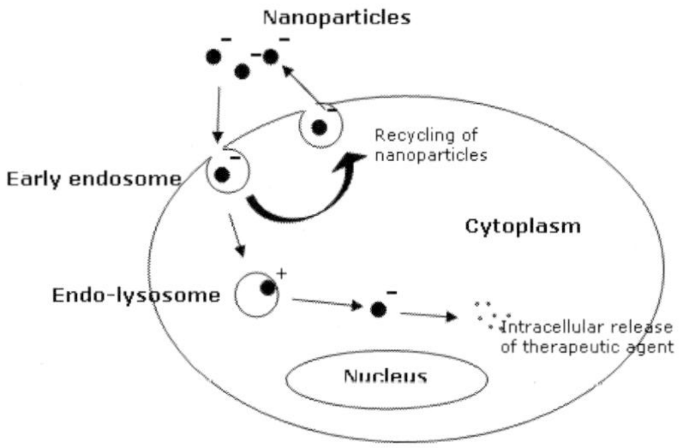

Fig. 2 Targeting of intracellular bacteria with antibiotic nanoparticles.

4. Therapeutic Applications

Recently, the use of polymeric nanoparticles as drug delivery devices has been extensively reviewed [Soppimath *et al.*, 2001], and nanoparticles have been mainly suggested for antibiotic targeting, sustained release for topical, oral and intravenous bioavailability improvement purposes [Lboutounne *et al.*, 2002; Smith, 2005; Moulari *et al.*, 2005, 2006; Nhung *et al.*, 2007].

4.1. Topical Administration

Several antibiotics are administrated by topical route. However, the frequent use of antibiotics and some scrubbing techniques can be the cause of skin irritation or allergies [Messager, 2001]. Then many therapeutics strategies, using nanoparticles, have been proposed on the one hand to improve the antibiotic bioavailability, and efficacy. And the other hand to limit the skin irritation or allergies caused by this administration type.

Chlorhexidine and its salts are widely used as antimicrobial agent; they are incorporated into hand-washing agents [Doebbeling *et al.*, 1992]. From the range of active chemical agent used for antisepsis, disinfection and preservation [McDonnell and Russell, 1999], chlrorhexidine which has broad-spectrum antibacterial activity against Gram positive and Gram negative bacteria [Aly and Maibach, 1979], has been considered the most acceptable for reducing nosocomial transmission of infections in intensive care units [Frantz *et al.*, 1997]. Therefore, the potential of chlrorhexidine incorporated either in liposomes or in microspheres to provide controlled delivery of the drug to skin-associated bacteria and sustained activity against Gram positive and Gram negative bacteria was investigated [Egbaria and Friedman, 1990; Jones *et al.*, 1997]. Liposomes, in spite of some advantages over conventional dosage forms, are limited in their use by low encapsulation efficiency, rapid leakage and poor storage stability [Gregoriadis, 1988]. This is why, others strategies nanoparticles-based have been developed.

Lboutounne *et al.* [2002], for example, investigated the potential of chlorhexidine base encapsulated into poly(ε-caprolactone) (PCL) nanocapsules by the in vitro bactericidal activity determination against several hospital bacterial strains. And the *ex vivo* antibacterial activity against *Staphylococcus epidermidis*, bacteria largely implicated in nosocomial infections, applied to porcine ear skin. The topical antibacterial efficacy of chlorhexidine base loaded PCL nanocapsules was compared to that obtained with a disinfectant-detergent solution of chlrorhexidine digluconate.

The results obtained have showed that chlorhexidine base loaded PCL nanocapsules maintained an antibacterial activity against several bacteria, during 40 days, suggesting a sustained release of chlorhexidine base from PCL nanocapsules (Table 1). Furthermore, a sustained release of chlorhexidine base from PCL nanocapsules enhanced drug delivery by mediating a more direct and prolonged contact between the carrier and bacteria, skin surface and skin follicles and presented a prolonged *ex vivo* topical antibacterial activity against *Staphylococcus epidermidis*. Always with chlorhexidine-encapsulated into PCL, the same effects have been obtained by Nhung *et al.* [2007] Indeed working on antimicrobial activity of chlrorhexidine, authors were compared, ex vivo, the sustained anti-microbial effect of chlrorhexidine-loaded nanocapsule-based gel (Nanochlorex® Pirot, 2006) against resident and transient skin flora with a commercial 62% (v/v) ethanol-based rub gel (Purell®).

Table 1 MICs of either 0.15% chlrorhexidine base loaded PCL nanocapsule after 1 and 40 days of storage 4°C, or 0.15% chlrorhexidine digluconate aqueous solution against same bacterial strains [Lboutounne *et al.*, 2002].

Bacterial strains	MIC (µg/ml)				
	0.15% Chlorhexidine base nanocapsules		0.15% Chlorhexidine digluconate	Literature data	
	Day 1	Day 40		Chlorhexidine [18,19]	MBC (µg/ml)
S. epidermidis	1.2	1.2	1.2	0.25–8	—
S. aureus CIP 53154	1.2	1.2	1.2	—	12.5–31.25–62.2
E. faecalis	9.4	Nd	4.7	1.4–3	—
E. hirae CIP 5855	1.2	1.2	9.4	—	125
E. aerogenes	37.5	37.5	75	2–64	—
P. aeruginosa	75	75	75	3–6	—
P. aeruginosa CIP A22	75	75	75	—	31.25–62.2
E. coli CIP 54127	1.2	1.2	1.2	—	15.6–31.25

Nd, not determined.

Eight informed volunteers were treated. The used method consisted to mount human skin samples onto vertical diffusion cells containing the saline solution in the receptor compartment. The formulation tested have been applied onto the delimited area skin in the donor compartment for 5 min. one millilitre of bacteria inocula was added to the skin surface. Authors evaluate the effectiveness of residual hand-gel, inocula were withdrawn at regular time intervals (1, 2, 3 and 4) and replaced by fresh bacterial suspension onto the skin surface (1,2 and 3h). Therefore, inocula collected from skin surface were diluted, and were subsequently incorporated into tryptic soy agar medium. The plates have been incubated for 48h at 35ºC, and the bacterial colonies were visually

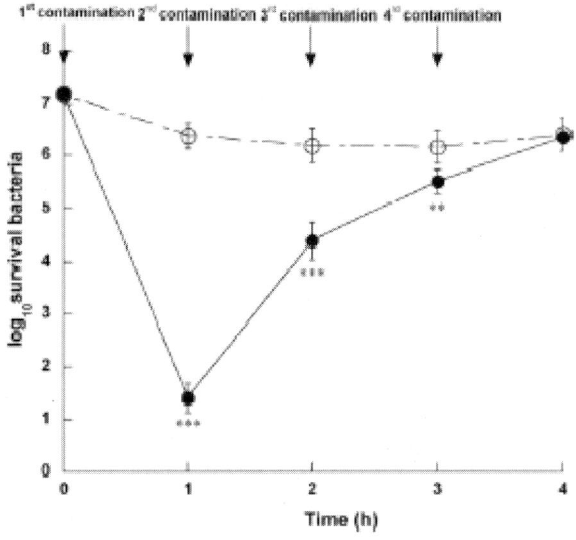

Fig. 3 Comparison of sustained antibacterial effect of (•) Nanochlorex® and (○) Purell® applied to human skin for 5 min, then contaminated successively by Staphylococcus epidermidis at time +5min, +1h and +3h. **p<0.01; ***p<0.0001 compared to Purell® group [Nhung *et al.,* 2007].

counted. The results obtained showed that only Nanochlorex® exhibited a sustained effect against bacteria (Figure 3).

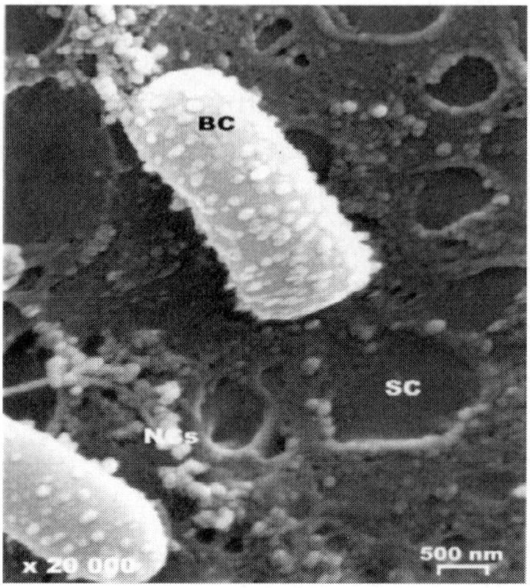

Fig. 4 Scanning electron micrographs of 0.60% chlorhexidine base loaded PCL nanocapsules localization on stratum corneum-associated bacteria. Drug loaded nanocapsules (NCs) adsorbed on bacteria membrane (BC) [Lboutounne *et al.*, 2002].

Moulari *et al.* [2005] working on plant extracts at antimicrobial activity shown that poly(lactide-co-glycolide) acid nanoparticles-loaded plant extract (500 and 1000 μg/ml) were presented, *ex vivo*, the best antibacterial activity against Staphylococcus epidermidis after the artificial contaminations of human skin samples than plant extracts solution at the same concentrations.

All these authors explain this antibacterial nanoparticles activity by the interactions of nanoparticles with bacteria components, which would facilitate the continuous diffusion of the antimicrobial agent from the core of nanoparticles in the bacterial membrane and limit the adhesion of bacteria on skin surface. Another phenomenon discussed was the bioadhesion phenomena of nanoparticles on bacterial membrane as shown it Lboutounne *et al.* [2002] (Figure 4).

4.2. Oral Administration

Oral delivery, in which the therapeutic agent is absorbed from gastrointestinal tract, is the most desirable approach, but success with antibiotics is limited by barriers to antibiotic absorption from the gastrointestinal tract. Nanoparticles can be used to protect a labile drug from degradation in the gastrointestinal tract and protect gastrointestinal tract from drug toxicity. Briefly, nanoparticles have been used as oral drug carriers for following reasons: improvement of the bioavailability of drugs with poor absorption characteristics [Couvreur *et al.*, 1980; Florence *et al.*, 1995], prolongation of the residence time of drugs in the intestine, high dispersion at the molecular level and consequently increase of absorption, control of the drugs release [Allémann *et al.*, 1992; Hubert *et al.*, 1991], Targeting of therapeutic agents to a particular organ and thus reducing toxicity [Espuelas *et al.*, 1997], reduction of the gastrointestinal mucosal irritation caused by drugs [Fessi et al., 1989; Ammoury *et al.*, 1991], assurance of the drugs stability in the gastrointestinal tract [Grangier *et al.*, 1991; Roques *et al.*, 1992].

Here we describe some examples of antibiotic drug that are being investigated for oral administration.

The oral route is often used to treat the gastric infections as well as the Helicobacter pylori infections. Indeed, H. pylori has become recognized as a major gastric pathogen with worldwide distribution. *H. pylori*, a prevalent human-specific pathogen, is a causative agent in chronic active gastritis [Warren and Marshall, 1983], gastric and duodenal ulcers [Megraud and Lamouliatte, 1992], and gastric adenocarcinoma [Forman *et al.*, 1994]. It is susceptible to many antibiotics in vitro but has proved difficult to eradicate in vivo. H. pylori penetrates the gastric mucus layer and fixes itself. Therefore, access of antimicrobial drugs to the site is restricted from both the lumen of the stomach and the gastric blood supply. H. pylori may also have acquired resistance to the commonly used antimicrobial agents. Because conventional antibiotics do not remain in the stomach for prolonged periods, they are unable to deliver the antibiotics to the infection site in effective concentrations and in fully active forms. Also, many antimicrobials agents, such as penicillin and erythromycin, degrade rapidly in an acidic environment.

It is therefore necessary to develop drug delivery systems able to optimize the antibiotics activity. Umamaheshwari *et al.* [2004], for example, investigated the anti-Helicobacter pylori of mucoadhesive nanoparticles bearing amoxicillin in experimental gerbils model.

Nanoparticles have been prepared with gliadin polymer, and their in vivo clearance study point out using gerbils. Six animals assigned to 3 groups, have been inoculated with 1 ml, via intragastric gavage, of H. pylori suspension. Fourteen days after infection, amoxicillin was orally administrated once a day for 3 consecutive days at a dose of 4, 10, or 40 mg/Kg in the nanoparticles or suspension form. The results obtained show that the mean bacterial count after oral administration of amoxicillin suspension decreased as the amoxicillin dose increased, but complete clearance of H. pylori has not obtained even with the highest dose. With nanoparticles-loaded amoxicillin (1 mg/kg), the mean bacterial count has been significantly lower than amoxicillin suspension. Complete clearance of H. pylori (clearance rate, 100%) has observed after the administration of nanoparticles at doses of 10 and 40 mg/Kg. These results demonstrate that nanoparticles provided 10 times greater anti H. pylori activity than the amoxicillin suspension.

Sharma *et al.* [2004] studied chemotherapy efficacy of poly(dl-lactide-co-glycolide) (PLG) nanoparticles encapsulated antitubercular drugs at sub-therapeutic dose against experimental tuberculosis. Indeed, the fact that a tuberculosis patient has to take multiple antitubercular drugs for at least 6 months is largely responsible for patient non-compliance and therapeutic failure. Thus, a reduction in dosing frequency is a welcome therapeutic strategy for a better management of tuberculosis. Authors have proposed to encapsulate, using PLG-nanoparticles, three front-line antitubercular antibiotics: rifampicin, isoniazid and pyrazinamide. And they have evaluated nanoparticles efficacy on a murine model.

Briefly, nanoparticles were prepared; the mice have been infected with *Mycobacterium tuberculosis* suspension ($1.5x10^5$ CFU/ml) through the intramuscular route [Pandey *et al.*, 2003 and 2005]. Twenty days post infection, the animals have been following divided: group 1 untreated controls; group 2 oral free drugs at therapeutic dose for 6 weeks, daily; group 3 oral free drugs at 2/3 rd therapeutic dose for 6 weeks, daily; group 4 oral drug loaded PLG-nanoparticle at 2/3 rd therapeutic dose every 10 days for 6 weeks (5 doses); group 5 oral drug loaded PLG-nanoparticle at therapeutic dose every 10 days for 6 weeks (5 doses); group 6 oral unloaded PLG-nanoparticle every 10 days for 6 weeks (5 doses). Nanoparticles as well as the free drugs have suspended daily in isotonic saline just before the oral administration. The animals have been sacrificed after 46 days of treatment, the spleen and caudal lobe of the right lung of each animal have removed and homogenized in isotonic saline solution. Tissue homogenates have plated on Middlebrook agar, colony forming units (CFU) have been enumerated 21 days post inoculation. The results obtained show the efficacy of 2/3 rd therapeutic dose of PLG-nanoparticle which was not equivalent to a therapeutic dose of PLG-nanoparticle but at the same time proved to be better than 2/3 rd therapeutic dose of free drugs.

By these results, we can say that nanoparticles offer patient a less heavy approach for the administration of anti-tuberculosis drugs bearing a high chemotherapeutic potential. The nanoparticles formulation had a notably increased bioavailability compared with that of the commercial formulation.

4.3. Intravenous Administration

Potential applications of colloidal drug carriers administered intravenously can be summarized in terms of the drugs concentration in accessible sites, the rerouting of drugs away from sites of toxicity, and increasing the circulation half-life of labile or rapidly eliminated drugs. Intravenous administration permits to treat the bacterial intracellular infections.

It is known that the intracellular location of many bacteria is obviously a privileged niche, well protected from the immune system and from the action of many antibiotics. The efficacy of the therapy of bacterial intracellular infection depends on an effective cooperation between host defences and antibiotics. Host defences are modulated by molecules such as cytokines. Antibiotics efficacy is governed by their intracellular penetration, accumulation, disposition and bioavailability. This is why, the need for antibiotics with greater intracellular efficacy led to the development of endocytosable drug carriers, such as nanoparticles, which mimic the entry path of the bacteria by penetrating the cells into phagosomes or lysosomes. Also, It is well known that particles uptake by phagocytic cells is generally affected by particle size and surface properties [Tabata and Ikada, 1991; O'Brien and Guidry, 1996; Müller *et al.*, 1997]. Efficient particle uptake requires particle diameters in the nano or low micrometer range, although larger particles can be phagocytosed to a lower extent. It was observed that the use of particulate carriers able to undergo endocytosis increased intracellular delivery of antibiotics [Stevenson *et al.*, 1983; Vladimrsky and Laddigina, 1982]. Thus, many studies showed that nanoparticles could enhance the activation of monocytes-macrophage. Indeed, upon particle phagocytosis, the mononuclear phagocytic system (MPS) increases the production of cytokines which are involved in host defence against pathogenic micro-organisms [Tabata and Ikada, 1991]. In addition, the inclusion of nanoparticles may result in the activation of macrophages and, subsequently, enhance host defence functions of the immune system [Artusson *et al.*, 1987]. Both the efficacy of nanoparticles uptake and monocytes activation may be of importance for the therapeutic efficacy of nanoparticles loaded with antibiotics to treat intracellular bacterial infections. A typical intracellular bacterial target for such delivery systems is *Brucella spp*, which mainly resides inside macrophages for long time periods thereby evading host defence mechanisms [Flesh and Kaufmann, 1990].

It is also known that the intravenously administration of nanoparticles presents the some obstacles. So, the reticuloendothelial system, mainly the liver and the spleen, is a major obstacle to active targeting because of its ability to recognize these systems, remove them from systemic circulation, and, consequently, avoid the effective delivery of the nanoparticles to organs other than those of the reticuloendothelial system [Kumar *et al.*, 2001]. However, surface modification of these nanoparticulate systems with hydrophilic polymers is the most common way to control the opsonization process and to improve the surface properties of the system, or coating modification with polymers [Soppimath *et al.*, 2001] such as the attachment of poly(ethylene glycol) (PEG) chains to biodegradable polymer such as poly(lactic acid) (PLA) and poly(lactic-co-glycolic acid) (PLGA). Thus the hydrophilic PEG chains allow the control of protein and peptide absorption and, in addition, will allow the regulation of cell behaviour at the polymer surface.

However, studies carried out to treat bacterial intracellular infections are made either on in vivo model, or on in vitro model (isolated cells).

4.3.1. In vitro Model

Fawaz *et al.*, [1998] investigated the efficacy of ciprofloxacin-loaded polyisobutylcyanoacrylate (PIBCA) nanoparticles against five strains *Mycobacterium avium* complex (MAC) in human macrophages isolated from AIDS patients compared with free drug solution. Four ciprofloxacin concentrations (4, 8, 16 and 32 μg/ml) in the incubation medium have been used. Macrophages obtained from peripheral blood of healthy donors. On day 7, macrophages have inoculated with a MAC suspension and incubated for 37°C in 5% CO_2 atmosphere. At days 0, 4 and 7, the medium from alternate plates was discarded and the monolayers were lysed by introducing the distilled water. Samples of supernatant and lysates of macrophages were taken for colonies forming units (CFU) counts at days of infection.

The results obtained show that with the association of ciprofloxacin-PIBCA nanoparticles the antimicrobial activity against *Mycobacterium avium* complex (MAC) in human macrophages was increased compared with drug solution. As the concentration of ciprofloxacin increased in the range 4-32 μg/ml the CFU reduction from both ciprofloxacin decreased with a better efficiency from nanoparticles.

Other in vitro studies have been realized by several authors. Fontana *et al.* [2001] prepared amoxicillin-loaded polyethylcyanoacrylate (PECA) nanoparticles in the presence and in the absence of PEG and they determined phagocytic uptake, in view of a possible use of this colloidal carrier for controlled intravenous drug delivery and targeting. The phagocytic uptake of nanoparticles was determined by using murine monocytes macrophage cell line J774 A.1. The results obtained were interesting compared with the free drug. In addition, the absorption of PEG on the nanoparticles surface could allow realizing stealth formulations which can ovoid mononuclear phagocytic system (MPS) recognition in vivo owing to the presence of hydrophilic groups of PEG chains.

Zhang *et al.* [1998] were showed that gentamicin–loaded poly(butylcyanoacrylate) nanoparticles presented efficient intracellular delivery of the antibiotic to infected cultured mouse peritoneal macrophages and rat hepatocytes.

Prior *et al.* [2002] showed the interest of particulate systems in the treatment of bacterial intracellular infections after studied having the phagocytic uptake of poly lactic acid (PLA)/poly(dl-lactide-co-glycolide) acid microspheres-loaded gentamicin on murine monocytes-macrophages model. In conclusion of their works, authors showed that incubation of PLA/PLGA microspheres with monocyte-macrophages resulted in particle uptake, however, opsonization by serum components affected differently the extent of phagocytosis of murine monocytes. High stimulation was induced by the nanoparticles. The results suggest that the microspheres developed may be useful for treating efficiently intracellular gentamicin-sensitive pathogens such as *Brucella spp.*

4.3.2. In vivo Model

Ampicillin is an antibiotic possessing a very short half-life (t1/2 = 1.3 ± 0.2 h) [Goodman and Gilman, 1990], so, for increase this half-life, Youssef *et al.* [1988] studied the effectiveness of nanoparticles-loaded ampicillin in the treatment of *Listeria monocytogenes* infection on murine model. Nanoparticles were prepared using polyisohexylcyanoacrylate polymer. L. monocytogenes were cultivated on blood agar at 37°C, the resulting culture was diluted and injected (0.2 ml) into each mouse through the tail vein. At 2, 5 and 8 days after bacterial contamination, the mice were treated by intravenously route with different formulations (nanoparticles and free drug). The mice were killed; their spleen and liver were immediately removed aseptically and homogenized in sterile saline. To count viable bacteria, authors plated each homogenate on blood agar and incubated them at 37°C for 48 h.

The results obtained by authors show that ampicillin-loaded polyisohexylcyano-acrylate nanoparticles present a better therapeutic effect than free ampicillin in the treatment of chronic listeriosis. Indeed, liver sterilization was observed by day 7 postinfection in the mice treated with nanoparticles-loaded ampicillin, but never in those treated with free ampicillin, even at a total dose of 48 mg. Furthermore, for 8 days after the last therapeutic injection, nanoparticles-loaded ampicillin in a total dose of 2.4 mg continued to act more effectively than a total dose of 48 mg of free drug.

Fattal *et al.* [1989] also have realized the study on ampicillin by using a murine model. After having prepared polyhexylcyanoacrylate ampicillin loaded nanoparticles, authors showed that the entrapment of ampicillin in polyhexylcyanoacrylate nanoparticles was found to increase by 120-fold the efficacy of the antibiotic in experimental murine salmonellosis. While the untreated animals died in 10 days, a survival of 100 % was obtained after administration of a single dose of 0.8 mg or a repeated dose of 3 x 0.8 mg of ampicillin – loaded nanoparticles. The authors attributed this result to additional effects. Firstly, the distribution studies showed that the drug was mainly concentrated in the liver and the spleen, organs with major foci of infection. Secondly, the cellular uptake of the nanoparticles is assumed to be increased.

Therapeutic applications of nanoparticles administered by intravenously route are promising and present a particular interest in the antibiotics targeting for the treatment of intramacrophagic opportunistic infections.

4.4. Other Routes

Nanoparticles have been used to encapsulate isoniazid, pyrazinamide and rifampicin for be administered by aerosol route on animal model (rats) [Zahoor *et al.*, 2005]. Nanoparticles were prepared using alginate polymer and nebulised by a compressor nebuliser system. The drugs were used at therapeutic dosage (isoniazid, 10 mg/Kg; pyrazinamide, 15 mg/Kg; rifampicin, 12 mg/Kg. animals were infected intramusculary with 1.5 x 105 viable bacilli of M. tuberculosis H37Rv in 0.1 ml of 0.9% NaCl. Infection was confirmed by the presence of mycobacteria in tissue smears of spleens and lungs of two to three animals after 20 days. The different animal groups received the drug doses. Animals were sacrificed after 45 days of chemotherapy. Lungs and spleen were removed aseptically and homogenized, and then the tissue homogenates was plated on culture medium, and colony forming units (CFUs) were enumerated 30 days post inoculation.

Table 2 Chemotherapeutic efficacy of aerolised alginate nanoparticle encapsulating antibercular drugs against experimental tuberculosis in guinea pigs [Zahoor *et al.*, 2005]

Group	$\log_{10}$ CFUs[a]	
	Lung (right caudal lobe)	Spleen (whole organ)
Untreated controls	5.8 ± 0.1	5.9 ± 0.1
Empty alginate nanoparticles every 15 days, aerosol (three doses)	$5.8 \pm 0.3^{*}$	5.9 ± 0.1
Drug-loaded alginate nanoparticles every 15 days, aerosol (three doses)	$<1.0^{b}$	$<1.0^{b}$
Free drugs daily, orally (45 doses)	$<1.0^{b}$	$<1.0^{b}$

CFUs, colony-forming units.

[a] Results are based on visible growth of *Mycobacterium tuberculosis* on Middlebrook 7H10 agar on day 21 post inoculation. Results are mean ± S.D. ($N = 5$–6).

[b] Value <1.0 indicates no detectable CFUs following the inoculation of 50 μL of neat and 1:10 diluted tissue homogenates.

[*] $P > 0.05$ according to ANOVA.

The results obtained show the detection of active principles in tissue what demonstrate that alginate nanoparticles encapsulating isoniazid, pyrazinamide and rifampicin, when administered through the aerosol route, not only increased local (lung) drug bioavailability but also the bioavailability of each drug at other sites of the body. Authors also observed that only three nebulised doses of this formulation, administered every 15 days, were equiefficient to 45 doses of oral free drugs administered daily in achieving undetectable CFUs (Table 2)

Nanoparticles have shown promising results over the last 10 years in ophthalmology, providing protection of drug from chemical and enzymatic degradation, improved tolerance, increased corneal uptake, and longer intraocular half-life [Lallemand *et al.*, 2003]. Subsequently, various types of nanoparticles were proposed to take advantage of prolonged residence time, because the short elimination half-life of ophthalmologic drugs remains a major problem in ocular therapy [Couvreur *et al.*, 1995].

Several other studies concerning the important role of nanoparticles-loaded antibiotic in the treatment of bacterial infections showed promising results such as the work of Forestier *et al.* [1992]. These searchers have shown the nanoparticles-bound ampicillin efficacy on the survival of *Listeria monocytogenes* in mouse peritoneal macrophages. Kelly *et al.* [2004] on incorporation of tetracycline into halloysite polymer for the treatment of periodontis, Alt *et al.* [2004], Sondi *et al.* [2004], Moulari *et al.* [2006] on plant extract, Qi *et al.* [2004], Beyth *et al.* [2006], Grace and Pandian [2007].

Many reviews and articles have highlighted the use of biodegradable polyesters and biocompatible inorganic materials [Trikeriotis and Ghanotakis, 2007] as effective drug carriers including nanoparticles or microparticles, hydrogels, micelles and fibrous scaffolds [Freiberg and Zhu, 2004; Varde and Pack, 2004; Moses *et al.*, 2003; Sinha and Trehan, 2003; Zhao *et al.*, 2003; Rabinow, 2004] in the treatment of different diseases such as the bacterial infections. Inevitably, there are advantages and limits associated with each drug delivery system.

4.5. Advantages and Limits of Antimicrobial Therapy by Nanoparticles

Nanoparticles-based drug delivery systems have considerable potential for the treatment of many bacterial infections. The important technological advantages of nanoparticles used as drug carriers are high stability, high carrier capacity, feasibility of incorporation for both hydrophilic and hydrophobic substances, and feasibility for variable routes of administration, including oral, intravenous, topic applications and inhalation. Nanoparticles can also be designed to allow controlled (sustained) drug release from the matrix. These properties of nanoparticles enable improvement of drug bioavailability, reduction of the dosage frequency, and may resolve the problem of toxicity such as hepatotoxicity and other side effects of many drug (antituberculosis drug for example), nephrotoxicity of aminoglycosides and various side effects of other antibiotic classes. However, the major limiting factor for the nanoparticulate technology application remains the big financial investment which it needs.

5. Conclusion

In this chapter we described amongst other things the limits of classical antibiotics, and showed the important role of the nanoparticles in the antibiotic therapy. Indeed colloidal drug carriers such as nanoparticles are able to modify the distribution of an associated substance. They can therefore be used to improve the therapeutic index of drugs by increasing their efficacy and/or reducing their toxicity.

Numerous studies showed that the resistance to antibacterial agent is increased [Gilbert *et al.*, 1990; Eng *et al.*, 1995], when bacteria stick to and embed in the glycocalix matrix. In order to eliminate persisting bacteria, either in an inaccessible site or in a state dormancy, new strategies must be developed for testing antibacterial agents in non-conventional conditions [Anwar *et al.*, 1990; Desnottes, 1995]. And the nanoparticulate technology is a promising approach for the improvement of the antibiotic treatment. The combination of different types of carriers could also represent a more rational design for improving antibacterial therapy.

In conclusion, there is a wealth of experimental data favouring the antibiotic delivery by nanoparticles which represent a tool to transport essential drugs across the blood-brain barrier that normally are unable to cross this barrier. The nanoparticles may be especially helpful for the treatment of bacterial infections including the intracellular infections. However, the mechanism of the nanoparticles-mediated transport of the antibiotics across the BBB at present is not fully elucidated. The most likely mechanism seems to be endocytosis by the endothelial cells lining the brain blood capillaries.

It is clear from this brief chapter of the current literature that the full spectrum of modern drug delivery approaches is being applied to antimicrobial therapy, but the prospects for a revolution and popularization in treatment seem some way distant. The problem, as Smith *et al.* [2005] notes, is not with the drug delivery system, rather it is their payload.

References

Allémann, E., Gurny, R. and Doelker, E. (1992). Preparation of aqueous polymeric nanodispersions by a reversible salting-out process: influence of process parameters on particle size. *Int. J. Pharm.,* 87, pp. 247-253.

Alt, V., Bechert, T., Steinrücke, P., Wagener, M., Seidel, P., Dingeldein, E., Domann, E. and Schnettler, R. (2004). An in vitro assessment of the antibacterial properties and cytotoxicity of nanoparticulate sylver bone cement. *Biomaterials*, 25, pp. 4383-4391.

Aly, R. and Maibach, H. I. (1979). Comparative study on the antimicrobial effect of 0.5% chlrorhexidine gluconate and 70% isopropyl alcohol on the normal flora of hands. *Appl. Environ. Microb.*, 37, pp. 610-613.

Ammoury, N., Fessi, H., Devissaguet, J. P., Dubrasquet and M., Benita, S. (1991). Jejunal absorption, pharmacological activity, and pharmacokinetic evaluation of indomethacin-loaded poly(D,L-lactide) and poly-isobutyl-cyanoacrylate) nanocapsules in rats. *Pharm. Res.*, 8, pp. 101-105.

Anwar, H., Dasgupta, M. K. and Costerton, J. W. (1990). Testing the susceptibility of bacteria in biofilms to antibacterial agents. *Antimicrob. Agents Chemother.*, 34, pp. 2043-2046.

Artusson, P., Arro, E., Edman, P., Ericsson, J. L. E. and Sjöholm, I. (1987). Biodegradable microspheres V: Stimulation of macrophages with microparticles made of various polysaccharides. *J. Pharm. Sci.*, 76, pp. 127-133.

Beyth, N., Yudovin-Farber, I., Bahir, R., Domb, A. and Weiss, E. (2006). Antibacterial activity of dental composites containing quaternary ammonium polyethylenimine nanoparticles against Streptococcus mutans. *Biomaterials*, 27, pp. 3995-4002.

Couvreur, P., Dubernet, C. and Puisseux, F. (1995). Controlled drug delivery with nanoparticles: current possibilities and future trends. *Eur. J. Pharm. Biopharm.*, 41, pp. 2-13.

Couvreur, P., Lenaerts, V., Kante, B., Roland, M. and Speiser, P. P. (1980). Oral and parenteral administration of insulin associated to hydrolysable nanoparticles. *Acta Pharm. Technol.*, 26, pp. 220-222.

Doebbeling, B. N., Stanley, G. L., Sheetz, C. T., Pfaller, M. A., Houston, A. K., Annis, L., Li, N. and Wenzel, R. P. (1992). Comparative efficacy of alternative hand-washing agents in reducing nosocomial infections in intensive care units. *New Engl. J. Med.*, 327, pp. 88-93.

Desnottes, J. F. (1995). Recherche de nouveaux antibactériens. Evolution des méthodes d'évaluation microbiologiques. *Bull. Soc. Fr. Microbiol.* (HS), 10, pp. 37-44.

Egbaria, E. and Friedman, M. (1990). Sustained in vitro activity of human albumin microspheres containing chlrorhexidine dihydrochloride against bacteria from cultures of organisms that cause urinary tract infections. *Antimicrob. Agents Chemother.*, 34, pp. 2118-2121.

Eng, R. H. K., Hsieh, A. and Smith, S. M. (1995). Antibiotic killing of bacteria: comparaison of bacteria on surfaces and in liquid, growing and non-growing. *Chemotherapy*, 41, pp. 113-120.

Espuelas, M. S., Legrand, P., Irache, J. M., Gamazo, C., Orecchioni, A. M. and Devissaguet, J. P. (1997). Poly(ε-caprolactone) nanoparticles as an alternative way to reduce amphotericine B toxicity. *Int. J. Pharm.*, 158, pp. 19-27.

Fawaz, F., Bonini, F., Maugein, J. and Lagueny, A. M. (1998). Ciprofloxacin-loaded polyisobutylcyanoacrylate nanoparticles: pharmacokinetics and in vitro antimicrobial activity. *Int. J. Pharm.*, 168, pp. 255-259.

Fattal, E., Youssef, M. and Couvreur, P., Andremont, A. (1989). Treatment of experimental salmonellosis in mice with ampicillin-bound nanoparticles. *Antimicrob. Agents Chemother.*, 33, pp. 1540-1543.

Fessi, H., Puisieux, F., Devissaguet, J. P., Ammoury, N. and Benita, S. (1989). Nanocapsule formation by interfacial deposition following solvent displacement. *Int. J. Pharm.*, 55, pp. R1-R4.

Flesh, I. E. A and Kaufmann, S. H. E. (1990). Cell-mediated immunity against intracellular bacteria. In: Adams, LG (Ed), Advances in Brucellosis research. Texas A and M University press, Texas, pp. 121-122.

Florence, A. T., Hillery, A. M., Hussain, N. and Jani, P. U. (1995). Nanoparticles as carriers for oral peptide absorption: studies on particle uptake and fate. *J. Control. Rel.*, 36, pp. 39-46.

Fontana, G., Licciardi, M., Mansueto, S., Schillaci, D. and Giammona, G. (2001). Amoxicillin-loadedpolyethylcyanoacrylate nanoparticles: Influence of PEG coating on the particle size, drug release rate and phagocytic uptake. *Biomaterials*, 22, pp. 2857-2865.

Forestier, F., Gerrier, P., Chaumanrd, C., Quero, A. M., Couvreur, P. and Labarre, C. (1992). Effect of nanoparticle-bound ampicillin on the survival of Listeria monocytogenes in mouse peritoneal macrophages. *J. Antimicrob. Chemother.*, 30, pp. 173-179.

Forman, D., Webb, P. and Parsonnet, J. (1994). H. pylori and gastric cancer. *Lancet*, 343, pp. 243-244.

Frantz, S. W., Haines, K. A., Azar, C. G., Ward, J. I., Homan, S. M. and Roberts, R. B. (1997). Chlorhexidine gluconate (CHG) activity against clinical isolates of vancomycin-resistant Enterococcus faecium and the effects of moisturizing agents on CHG residue accumulation on the skin. *J. Hosp. Infect.*, 37, pp. 157-164.

Freiberg, S. and Zhu, X. X. (2004). Polymer microspheres for controlled drug release. *Int. J. Pharm.*, 282, pp. 513-521.

Gilbert, P., Collier, P. J. and Brown, M. R. W. (1990). Influence of growth rate on susceptibility to antimicrobial agents: biofilms, cell cycle, dormancy and stringent response. *Antimicrob. Agents Chemother.*, 34, pp. 1865-8.

Goodman, L. S. and Gilman, A. (1990). The pharmacological basis of therapeutics, 8 th edn. New York: Pergamon Press Inc., pp. 1550-1627.

Grace, N., A. and Pandian, K. (2007). Antibacterial efficacy of aminoglycosidic antibiotics protected gold nanoparticles: A brief study. Colloids and Surfaces A: *Physicochem. Engineer Aspects*, 297, pp. 63-70.

Grangier, J. L., Puygrenier, M., Gauthier, J. C. and Couvreur, P. (1991). Nanoparticles as carriers for growth hormone releasing factors (GRF). *J. Control. Rel.*, 15, pp. 3-13.

Gregoriadis, G. (1988). Liposomes as drug carriers. Wiley, New York, London, Sydney, Toronto.

Grislain, L., Couvreur, P., Lenaerts, V., Roland, M., Deprez- Decampeneere and A., Speiser, P. (1983). Pharmacokinetics and distribution of a biodegradable drug carrier. *Int. J. Pharm.*, 15, pp. 335-45.

Hubert, B., Atkinson, J., Guerret, M., Hoffman, M., Devissaguet, J. P. and Maincent, P. (1991). The preparation and acute antihypertensive effects of a nanocapsular form of darodipine, a dihydropyridine calcium entry blocker. *Pharm. Res.*, 8, pp. 734-738.

Jones, M. N., Song, Y. H., Kaszuba, M. and Reboiras, M. D. (1997). The interaction of phospholipids liposomes with bacteria and their use in the delivery bactericides. *J. Drug Targeting*, 5, pp. 25-34.

Kaprelyants, A. S., Gottschal, J. C. and Kell, D. B. (1993). Dormancy in non-sporulating bacteria. *FEMS Microbiol. Reviews*, 104, pp. 271-286.

Kelly, H. M., Deasy, P. B., Ziaka, E. and Claffey, N. (2004). Formulation and preliminary in vivo dog studies of a novel drug delivery system for the treatment of periodontitis. *Int. J. Pharm.*, 274, pp. 167-183.

Kumar, N., Ravikumar, M. N. V. and Domb, A. J. (2001). Biodegradable block copolymers. *Adv. Drug Deliv. Rev.*, 53, pp. 23-44.

Lallemend, F., Felt-Baeyens, O., Besseghir, K., Behar-Cohenc, F. and Gurny, R. (2003). Cyclosporine a delivery to the eye: a pharmaceutical challenge. *Eur. J. Pharm. Biopharm.*, 56, pp. 307-318.

Lboutounne, H., Chaulet, J. F., Ploton, C., Falson, F. and Pirot, F. (2002). Sustained ex vivo skin antiseptic activity of chlorhexidine in PCL nanocapsule encapsulated form and as a digluconate. *J. Control. Release*, 82, pp. 319-334.

Li, Y., Leung, P., Yao, L., Song, Q. W, and Newton, E. (2006). Antimicrobial effect of surgical masks coated with nanoparticles. *J. Hosp. Infect.*, 62, pp. 58-63.

Mc Donnell, G. and Russel, A. D. (1999). Antiseptics and disinfectants: activity, action, and resistance. *Clin. Microbiol. Rev.*, 12, pp. 147-179.

Megraud, F. and Lamouliatte, H. (1992). Helicobacter pylori and duodenal ulcers. *Dig. Dis. Sci.*, 37, pp. 769-772.

Messager, S., Goddard, P. A., Dettmar, P. W. and Maillard, J. Y. (2001). Determination of the antibacterial efficacy of several antiseptics tested on skin by an ex vivo test. *J. Med. Microbiol.*, 50, pp. 284-292.

Moses, M. A., Brem, H. and Langer, R. (2003). Advancing the field of drug delivery: taking aim at cancer. *Cancer Cell*, 4, pp. 337-341.

Moulari, B., Lboutounne, H., Chaumont, J.-P., Guillaume, Y., Millet, J. and Pellequer, Y. (2006). Potentiation of the bactericidal activity of Harungana madagascariensis Lam. ex Poir (Hypericaceae) leaf extract against oral bacteria using PLGA-nanoparticles: in vitro study. *Acta Odontol. Scand.*, 64, pp. 153-158.

Moulari, B., Lboutounne, H., Pellequer, Y., Guillaume, Y., Millet, J. and Pirot, F. (2005). Vectorization of Harungana madagascariensis Lam. ex Poir (Hypericaceae) ethanolic leaf extract by using PLG-nanoparticles: antibacterial assessment. *Drug Dev. Res.*, 65, pp. 26-33.

Müller, R. H., Rühl, D., Lück, M. and Paulke, B. R. (1997). Influence of fluorescent labelling of polystyrene particles on phagocytic uptake, surface hydrophobicity, and plasma protein adsorption. *Pharm. Res.*, 14, pp. 18-24.

Nhung, D., Freydiere, A.-M., Constant, H., Falson, F. and Pirot, F. (2007). Sustained antibacterial effect of a hang rub gel incorporating chlrorhexidine-loaded nanocapsules (Nanochlorex®). *Int. J. Pharm.*, 334, 166-172.

O'Brien, C. N. and Guidry, A. J. (1996). Formulation of poly(D,L-lactide-co-glycolide) microspheres and their ingestion by bovine leukocytes. *J. Dairy Sci.*, 79: 1954-1959.

Pandey, R., Zahoor, A., Sharma, S. and Khuller, G. K. (2003). Nanoparticle encapsulated antitubercular drugs as a potential oral drug delivery system against murine tuberculosis. *Tuberculosis*, 83, pp. 373-378.

Pandey, R., Sharma, S. and Khuller, G. K. (2005). Oral solid lipid nanoparticle-based antitubercular chemotherapy. *Tuberculosis*, 85, pp. 415-420.

Poste, G (1983). Liposomes targeting in vivo: problems and opportunities. *Biol. Cell*, 47, pp. 19-39.

Prior, S., Gander, B., Blarer, N., Merkle, H. P., Subira, M. L., Irache, J. M. and Gamazo, C. (2002). In vitro phagocytosis and monocytes-macrophage activation with poly(lactide) and poly(lactide-co-glycolide) microspheres. *Eur. J. Pharm. Sci.*, 15, pp. 197-207.

Qi, L., Xu, Z., Jiang, X., Hu, C. and Zou, X. (2004). Preparation and antibacterial activity of chitosan nanoparticles. *Carbohydrates Res*, 339, pp. 2693-2700.

Rabinow, B. E. (2004). Nanosuspensions in drug delivery. *Nat. Rev. Drug Discov.*, 3, pp. 785-796.

Roques, M., Damgé, C., Michel, C., Staedel, C., Crémel G. and Hubert, P. (1992). Encapsulation of insulin for oral administration preserves interaction of the hormone with its receptor in vitro. *Diabetes*, 41, pp. 451-456.

Sharma, A., Pandey, R., Sharma, S. and Khuller G. K. (2004). Chemotherapeutic efficacy of poly(D,L-lactide-co-glycolide) nanoparticle encapsulated antitubercular drugs at sub-therapeutic dose against experimental tuberculosis. *Int. J. Antimicrob. Agents*, 24, pp. 599-604.

Sinha, V. R. and Trehan, A. (2003). Biodegradable microspheres for protein delivery. *J. Control. Rel.*, 90, pp. 261-280.

Smith, A. W. (2005). Biofilms and antibiotic therapy: Is there a role for combating bacterial resistance by the use of novel drug delivery systems. *Adv. Drug Deliv. Rev.*, 57, pp. 1539-1550.

Sondi, I. S. and Salopek-Sondi, B. (2004). Silver nanoparticles agent: a case study E. coli as a model for Gram-negative bacteria. *J. Colloid Interface Sci.*, 275, pp. 177-182.

Soppimath, K. S., Aminabhavi, T. M., Kulkarni, A. R. and Rudzinski, W. E. (2001). Biodegradable polymeric nanoparticles as drug delivery devices. *J. Control. Rel.*, 70, pp. 1-20.

Stevenson, M., Baillie, A. J. and Richards, R. M. E. (1983). Enhanced activity of streptomycin and chloramphenicol against intracellular Escherichia coli in the J774 macrophage cell line mediated by liposome delivery. *Antimicrob. Agents Chemother.*, 24, pp. 742-749.

Tabata, Y. and Ikada, Y. (1991). Phagocytosis of polymeric microspheres. In: Szycher, M (Ed.), High Performance Biomaterials. Technomic Publishing, Lancaster, pp. 621-646.

Trikeriotis, M. and Ghanotakis, D. F. (2007). Intercalation of hydrophilic and hydrophobic antibiotics in layered double hydroxides. *Int. J. Pharm.*, 332, pp. 176-184.

Turos, E., Shim, J.-Y., Wang, Y., Greenhalgh, K., Reddy, G., Dickey, S. and Lim, D. (2007). Antibiotic-conjugated polyacrylate nanoparticles: New opportunities for development of anti-MRSA agents. *Bioorg. Med. Chem. Letters*, 17, pp. 53-56.

Umamaheshwari, R. B., Ramteke, S. and Jain, N. K. (2004). Anti-Helicobacter pylori effect of mucoadhesive nanoparticles bearing amoxicillin en experimental gerbils model. *AAPS Pharm. Sci. Tech.*, 5, pp. 1-9.

Varde, N. K., Pack, D. W. (2004). Microspheres for controlled drug delivery, *Expert Opin. Biol. Ther.* 4, pp. 35-51.

Vladimrsky, M. A. and Ladigina, G. A. (1982). Antibacterial activity of liposome-entrapped streptomycin in mice infected with Mycobacterium tuberculosis. *Biomed. Pharmacother.*, 36, pp. 375-377.

Warren, J. R. and Marshall, B. (1983). Unified curved bacilli on gastric epithelium in active chronic gastritis. *Lancet*, 321, pp. 1273-1275.

Youssef, M., Fattal, E., Alonso, M. J., Roblot-Treupel, L., Sauzières, J., Tancrède, C., Omnès, A., Couvreur, P. and Andremont, A. (1988). Effectiveness of nanoparticle-bound ampicillin in the treatment of Listeria monocytogenes infection in athymic nude mice. *Antimicrob. Agents Chemother.*, 32, pp. 1204-1207.

Zahoor, A., Sharma, S. and Khuller, G. K. (2005). Inhalable alginate nanoparticles as antitubercular drug carriers against experimental tuberculosis. *Int. J. Antimicrob. Agents*, 26, pp. 298-303.

Zhang, Q., Liao, G., Wei, D., and Nagai, T. (1998). Increase in gentamicin uptake by cultured mouse peritoneal macrophages and rat hepatocytes by its binding to poly(butylcyanoacrylate) nanoparticles. *Int. J. Pharm.*, 164, pp. 21-27.

Zhao, Z., Wang, J., Mao, H. Q., and Leong, K. W. (2003). Polyphosphoesters in drug and gene delivery. *Adv. Drug Deliv. Rev.*, 55, pp. 483-499.

Chapter 8

NANOPARTICLE THERAPY IN PARASITES DISEASE: POSSIBILITY AND REALITY!

Malika Larabi

Stanford University School of Medicine, Lucas MRS Imaging Center, Stanford, California (USA)

1. Introduction

Infectious parasites are rapidly becoming a critical world public health issue. The past three decades have seen an increase in parasite resistance to conventional therapy, and a pronounced increase of immunodeficient patients as Human Immunodeficiency Virus (HIV). Furthermore, for some diseases a safe cure does not exist. Too often epidemics are localized in poor or developing nations where unstable government infrastructure or environmental catastrophe makes it incredibly difficult to motivate major pharmaceutical corporations to develop viable treatment. Current therapies are rapidly becoming obsolete; of the drugs used today to combat parasitic disease, most were developed during the first half of the twentieth century. These were developed in response to veterinary needs and the effects of colonialism, since wars often fermented in regions where parasitic disease was endemic.

Today's medical landscape is sobering. Of the 1393 new pharmaceuticals marketed between 1975 and 1999, only 13 were for parasitic disease of *any kind* [Trouiller *et al.*, 2002]. Furthermore, few vaccines against parasitic disease even exist. When western nations *do* find an avenue by which to utilize expensive drug therapies to treat opportunistic parasitic diseases, hundreds of thousands of patients are cured. But the overwhelming reality of the situation is that in any given year, these diseases will ultimately cause more 10 million deaths and make invalids of 20 million other humans across the globe.

There is, nevertheless, reason for hope. During the last five years, in front of the inertia of big pharmaceutical corporations and the overworked of World Health Organization (WHO), despite the colossal recrudescence of many diseases, many private and public partnerships have been formed. In collaboration with researchers who live where disease is endemic, these partnerships have mobilized efforts to develop drugs to effectively combat parasitic disease. Meanwhile, other programs and organizations screen drug libraries within pharmaceutical corporations and/or academia in the hopes of wringing new applications from old drugs. All of these components work toward the same goal; the reduction of parasite diseases via improved therapy and vaccines.

High-throughput screening technologies present an efficient way to develop new, powerful pharmaceuticals. Nevertheless, in recent years it has become evident that progress in drug development and therapy alone cannot make sufficient headway in the war against parasites. Poor water solubility of drug molecules, insufficient bioavailability, fluctuating plasma levels, and high food dependency are common barriers to success. As a result, efforts are underway to develop customized drug carriers that can overcome the current disappointing *in vivo* fates that many drugs ultimately face. Researchers hope that nanosized carriers can increase the solubility, and therefore the bioavailability, of active ingredients belonging to classes II and IV of the Biopharmaceutical Classification System.

The benefits of nanoparticle therapy extend beyond solubility. Nanoparticles can reduce toxicity more effectively reach target areas, and improve pharmacodynamic and pharmacokinetics of drugs whose effectiveness has been compromised by parasite recrudescence, such AmBisome in *Leishmania* diseases. Given the exponential rate of development within the field of nanotechnology, within mere years the cost to produce these nanoparticles should considerably diminish. As a result, their role in parasitic disease therapy will become increasingly critical. These wider applications cannot be overlooked; the prevention and the treatment of the entire host (human and other animal) are necessary for the eradication of these diseases.

After a brief introduction to the most common parasitic diseases, we will explore the dynamics of different therapeutic nanoparticles *in vivo*, possible complications in clinical trials, and their potential applications following approval.

2. Parasite Diseases

2.1. The Amoeba

Humans are infected by four main genera of amoeba, *Entamoeba, Endolimax,* Iodamoeba and Dientamoeba. These amoeba prefer to inhabit the intestinal tract of their chosen vertebrate or invertebrate hosts. Surprisingly, Entamoeba histolytica is the only species found to be associated with intestinal disease. Though many people worldwide harbor this organism, only about 10% develop clinically invasive disease. The pathogenic or invasive form invades the intestinal mucosa, produces dysentery, and in some circumstances gives rise to extra-intestinal lesions via the blood. This results in the classic symptoms of amoebiasis, which includes fulminating diarrhea, stomach cramps and vomiting. If left untreated, liver abscesses destroy hepatic tissue causing the death of the host. The host's strong humoral response does not provide adequate protection. Upper quadrant pain, fever, and weight loss are symptomatic of the acute form of this phase of the disease. Stool examination and Enzyme-Linked Immunosorbant Assays (ELISA) provide the principal methods of diagnosis.

2.2. Schistosomiasis

The disease is found in tropical countries such as Africa, the Caribbean, eastern South America, in east Asia, and in the Middle East and the particular area of the body in which the parasite localizes also determines its particular clinical profile. Though it has a low mortality rate, schistosomiasis can be highly debilitating. A victim under the throes of Katayama's fever will suffer abdominal pain, coughing, diarrhea, eosinophilia, fever, fatigue and hepatosplenomegaly. In the most severe cases, central nervous system lesions develop. If ectopic *S. japonicum* eggs form in the brain, cerebral granulomatous disease may result.

While praziquantel is both safe and effective in curing infected patients, it does not prevent re-infection by cercariae, and thus is not an optimal treatment for people living in areas where Schistosoma proliferates. In the past, antimony has also been used as a treatment option. Oxamniquine treats Schistosoma mansoni exclusively. Mirazid, a promising effective Egyptian oral pharmaceutical is currently being investigated. Overall, there has been substantial progress in the battle against this parasite.

2.3. Echinococcosis

The disease results from infection by tapeworm larvae of the genus *Echinococcus*. Definitive hosts are usually carnivores, such as dogs. Intermediate hosts are usually herbivores such as sheep and cattle, although humans sometimes fit into this category.

When E. granulosus infects its intermediate host, cysts form in the liver, lungs, kidneys, and spleen. These large cysts put tremendous pressure on blood vessels and organs, which may rupture or erode. Without immediate surgical intervention, death quickly follows shock. During a full-blown infection, the diameter of some cysts can exceed the diameter of a soccer ball (cystic hydatid). If the patient does not tolerate benzimidazole well, the only medical treatment currently available, or if the treatment fails, no further options are available. Recently however, amphotericin B was used with some success in three patients who had advanced to the pulmonary stage of the disease [Reuter *et al.*, 2003].

2.4. Toxoplasmosis

Toxoplasmosis is a parasitic disease caused by the protozoa *Toxoplasma gondii*. It is estimated that between 30 to 60 percent of the world population harbors this organism. Transmission is most common through the ingestion of raw or undercooked meat. Infection rarely causes illness; however some immunocompetent adults may experience a *flu*-like illness. In most non-immunodeficient patients, they follow a latent phase, during which cysts form in nerve and muscle tissue. The parasite can cause encephalitis, various neurological diseases, and can also affect the heart, liver, and eyes. Congenital toxoplasmosis can cause damage to the brain or the eyes. Even in its acute form, healthy patients with normal immune systems rarely require treatment. Pregnant women stricken with acute toxoplasmosis are given a combination of spiramycin and pyrimethaminecan, or simply spiramycin alone. Unfortunately, drugs which treat acute toxoplasmosis are unable to reach cysts in peripheral tissues or the brain in adequate concentrations.

2.5. Cryptosporidiosis

Cryptosporidium oocysts have a ubiquitous geographic distribution and can be detected in most surface drinking water sources. Since human infection was first documented in 1976, *Cryptosporidium parvum* has been recognized worldwide as a major cause of human dysentery. Cryptosporidiosis is a critical cause of persistent diarrhea in countries where the disease is endemic. Chronic cryptosporidiosis may contribute to complications such as biliary tract disease and malabsorption. The low number of annual deaths does not adequately reveal the carnage that this parasite exacts on humanity. There is no specific treatment for cryptosporidiosis. As of 2002, only nitazoxanide has been approved for treatment of diarrhea caused by *Cryptosporidium parvum.*

2.6. Trypanosoma

Infection by *Trypanosoma* manifests as three different diseases depending on the particular strain. Colloquially, we know two of them as African sleeping sickness. A third manifestation, Chagas disease, is found in South America. *Trypanosoma brucei rhodesiense* causes acute illness in Eastern and Southern Africa. If untreated, trypanosomiasis is fatal. A host's natural defenses can usually destroy most of the invading parasites. But the antigenic variation allows a small number of parasites to escape annihilation by the immune system. In South America, a different trypanosome, *T. cruzi*, causes Chagas disease.

The disease progresses through two distinct stages. Shortly after infection, there is an acute stage, after which there is a period of latency that may last several years. A chronic stage of the disease develops;

during which lesions form that irreversibly damage internal organs that include the heart, esophagus, colon, and the peripheral nervous system.

Pentamidine is the recommended treatment for African trypanosomiasis. If the disease has affected the central nervous system, melarsoprol and eflornithine are required. Acute Chagas disease therapy includes benznidazole and nifurtimox, and only benznidazole for intermediate Chagas disease. Other drugs have been tested in clinical development programs; these include allopurinol and the antifungal triazoles.

2.7. Malaria

The Plasmodium parasite that causes Malaria is among the most discriminating of parasites. Malarial episodes can be severe and fatal. Unless the parasite is completely eradicated, episodes are certain to recur. Treatment is becoming increasingly challenging. Already we have begun to witness the rise of drug resistant *Plasmodium falciparum* strains. Over the past decade, new anti-malarial compounds, such as artesunate, artemether and dihydroartemisinin, have been deployed on an increasingly large scale. WHO recommends that all countries experiencing resistance to conventional mono-therapies such as chloroquine, amodiaquine or sulfadoxine–pyrimethamine, should instead turn to combination therapies, preferably those containing artemisinin derivatives specific to *falciparum* malaria.

2.8. Leishmaniasis

The phlebotomine sandfly is the prime vector for the spread of protozoan flagellates of the genus Leishmania. These parasites cause varied clinical forms of leishmaniasis. Visceral leishmaniasis is often fatal if untreated. Muco-cutaneous leishmaniasis is a mutilating disease. Diffuse cutaneous leishmaniasis is disabling, and cutaneous leishmaniasis can result in disfiguration if multiple lesions occur. After 60 years of use pentavalent antimonials, are being gradually phased out as the primary treatment option for visceral leishmaniasis due to the resistance. Pentamidine is also limited by the irresversible toxicity it induces (insulin dependant diabetes, mellitus, and death). Miltefosine, a new alkylphosphocholine registered for use in India in 2002, is the first oral drug for treatment for visceral leishmaniasis. Paramomycin, an aminoglycoside holdover from the 1960s, has also been recently approved for use in India. It has a potentially versatile array of therapeutic applications, and in association with other drugs, is effective against both cutaneous and visceral leishmaniasis. Three such lipid associated formulations of amphotericin are commercially available: (*i*) liposomal amphotericin B, AmBisome; (*ii*) amphotericin B lipid complex, Abelcet and (*iii*) amphotericin B colloidal dispersion, Amphocil. Unfortuntately, vaccines under investigation for both cutaneous and visceral leishmaniasis are not yet ready for long-term use.

If there one common thread links the preceding eight parasites, it is that they have a complex lifecycle that requires at least two hosts. The condition under which each thrives is uniquely determined by its life cycle and the pathologies it induces. Drugs that combat these parasites must be incredibly versatile. An effective drug must be able reach different locations in the body so as to minimize the probability of relapse. Ideally, each drug must be effective in different hosts (human, pet or farm animal). As a result, numerous targets must be taken into account during the drug formulation process. In the case of *C. parvum*, which causes extreme diarrhea, or when *Leishmania* persists in macrophages, drug delivery and efficacy can be significantly enhanced by using simple additives and devices.

Developing new drugs is only half the battle. It is equally critical to improve kinetics and specificity, thereby reducing the serious side effects of existing drugs. In this context, nanoparticle therapy is used to increase the therapeutics index of old and new drugs, to hinder parasites' ability to develop resistance, and for prophylaxis use.

3. Nanoparticle Therapy

For parenteral administration, a nanoparticle carrier will both increase the solubility and stability of a drug. It will also reduce toxicity and modify its pharmacokinetic parameters by prolonging circulation time and improving its tissue distribution. Nanoparticle carriers also have the potential to promote increased drug internalization by phagocyte cells of the reticulo-endocytosis system, which often are reservoirs of intracellular parasites or co-localized in the infection site. This would lead to increased efficacy in peripheral infection targeting sites.

3.1. Micelles and emulsions

The micellar carrier system was among the first developed, and it is commonly used in drug formulation to improve solubility or reduce the toxicity. The micelization of amphotericin B is an "academic case". It has been used in fungal infections and as a secondary antiparasitic drug since the 1960s. Fungizone, which consists of amphotericin B and sodium desoxycholate micelles, is the gold standard in systemic fungal infection treatment. However, over 30% of patients receiving high doses of Fungizone show signs of renal disorder.

Heat treatment of Fungizone resulted in a super-aggregated micelle that decreased the level of toxicity in mammalian cells and increased activity against Leishmania donovani in BALB/c mice [Gaboriau *et al.*, 1997; Petit *et al.*, 1999].

Other micelles developed during the past few decades have shown promising results without having to go through preclinical studies. Amphotericin B micelles with the poloxamer 188 have shown a reversion of the resistance of amphotericin B in vitro [Espuelas *et al.*, 2000]. During the 1990s, different types of micelles mixed from biocompatible polymers of amphotericin B, (poly(ethylene glycol)-b-poly(epsilon-caprolactone), and 1,2-distearoyl-sn-glycero-3-phosphoethanolamine-N-methoxy-poly(ethylene glycol) were made to improve *in vivo* circulation time [Vakil *et al.*, 2005]. Targeted carriers increased efficacy and reduced toxicity in *Leishmania* infected mice as shown by targeted micelles composed of N-(2-hydroxypropyl)methacrylamide copolymers containing N-acetylmannosamine of amphotericin B [Nan *et al.*, 2004]. However, few of the above mixtures have been subjected to clinical studies as compared to other mixed micelles of amphotericin B. The limited number of preclinical studies can be attributed to the potential risk of hemolytic activity, the occurrence of anaphylactic shock, and poor pharmacokinetics.

Topically administered pharmaceuticals also utilize micellar carriers. Clinical trials investigating paramomycine creams reveal major improvements in cutaneous leishmaniasis therapy across the old world [Ozgoztasi *et al.*, 1997]. An uptake study involving murine skin confirmed in-vitro skin penetration and retention. Several formulations were tested in-vivo in Leishmania major cutaneous lesions in BALB/c mice. Topical treatment of paromomycin (15%) and gentamycin (0.5%) healed all lesions caused by Leishmania panamensis or Leishmania amazonensis with no indication of relapse [Grogl *et al.*, 1999]. Some surfactant polymers have been used for interactions with the lipid membrane that can increase the absorption of the drug. For example, poloxamer 188 has been used in the oral micellar formulation of atovaquone (Mepron) against malaria and toxoplasmosis. In addition, the lipid-bile salt mixed micelles promote intestinal lymphatic absorption [Dangi *et al.*, 1998].

Ivermectin is another successful micellar carrier. Introduced in the mid-1980, ivermectin is quite possibly the most broad-spectrum anti-parasitic medication ever developed. Oral paste formulations of Ivermectin were 100% effective against *Oxyuris equi* larvae, while intravenous formulations were 93% effective [Torbert *et al.*, 1982].

Another route of administration that can effectively target infectious lung parasites is the pulmonary route. In most cases, pulmonary administration of amphotericin B demonstrates a blood concentration similar to Fungizone when administered intravenously. In addition, the circulation time and high concentration in the lung are extent in the pulmonary infected animal, due to the presence of a higher number of phagocyte cells that absorb amphotericin B and release it slowly [Beyer *et al.*, 1993]. These functional parameters are modified differently depending on the carrier being used (lipid/polymer/charge [Lambros *et al.*, 1997]).

In this field, nanotechnology can improve locally administered therapy through specific carriers that deepen and slow drug release.

Micelle systems are the simplest formulations and remain the first option for poorly soluble molecules. However, their use is still limited by specific the physical properties, stability, or toxicity of the surfactant.

Emulsions are among of the oldest colloidal systems used to improve drug delivery. When Amphotericin B is incorporated into different types of emulsions, in particular phosphatidylcholine, general improvements in leishmania treatment are observed. When other therapy options fail, or if AmBisome is not affordable, clinicians in epidemic zones have turned to emulsion based delivery systems.

We find however that these Fungizone-fat emulsion sizes are larger than conventional Fungizone. This implies an increased drug concentration in mononuclear phagocyte system organs and improved therapy for visceral infections like Leishmania. Various other amphotericin B-emulsions have been recently developed. These include lecithin based oil emulsion for oral and parenteral administration [Brime *et al.*, 2003], or with Myrj 59, polyethylene glycol, pluronic [Tasset *et al.*, 1991] or triglycerides [Souza *et al.*, 1993]. Soybean oil amphotericin B emulsions have shown long-term anti-leishmania efficacy in dogs, which remain the parasite's primary host reservoir in most of Southern Europe [Cortadellas, 2003]. Each of these studies confirmed that enhanced efficacy suppressed relapse through improved intracellular therapy and distribution in peripheral tissue. By the same method, antimalarial activity of chloroquine improves after intra-peritoneal injection of Intralipid® or Ivelip®. Different strategies can improve the efficacy and stability of these emulsions. The charge of the carrier can significantly influence antiparasitic activity. Electric-charge inducers have demonstrated antimalarial activity in micellar oil-in-water emulsion formulations. Examples of this are mefloquine and halofantrine. Positively charged stearylamine exhibits a higher chemical affinity for cell membranes that leads to improved drug efficacy [Mbela *et al.*, 1998]. There was significant reduction of toxicity following intravenous injection of piperine loaded stealth lipid emulsions [Veerareddy *et al.*, 2004].

These emulsions are affordable and useful for parenteral administration. Moreover, they are promising candidates for oral delivery meant to target the lymphatic bypass, and for local applications such as creams. Micelles and emulsions have improved the pharmacokinetics, efficacy, and tolerance threshold these antiparasitic drugs. They are still commonly used colloidal formulations. However, they are limited by the stability, the faster intestinal transit, and the inefficient body distribution of the drug and its difficulty in targeting the specific infectious site.

3.2. Complexes

Intestinal absorption of meglumine is increased by its complexation with hydroxypropyl-beta-cyclodextrine. This led to improved efficacy in Leishmania amazonensis mice models. Infection parasite loads were diminished in cutaneous lesions [Demicheli *et al.*, 2004]. Modifying drug formulations as complexes with cyclodextrin improves also the drug efficacy on the migrating parasite stage and modifies the pharmacokinetic properties of albendazole. Following proven in vivo efficacy in goats and ovine, different antiparasitic drugs were tested in twelve, healthy male volunteers. Tests compared the bioavailability of β- and γ-cyclodextrin artemisinin complexes with normal, commercially available preparation Artemisinin250 (250 mg). Bioavailability and pharmacokinetic parameter values significantly differed between the complexes when half of the dose of the artemisinin-cyclodextrin complexes was compared to Artemisinin250. There was no significant difference between the values of the β- and γ-artemisinin-cyclodextrin complexes. These findings indicate that the β- and γ-cyclodextrin complexes have extended of bioavailability not only the anticryptosporidial drug has been improved by them complexation but also the study of the mechanism involved shown than these cyclodextrins by themselves have an anticryptosporidial activity [Castro-Hermida *et al.*, 2001; 2003; 2004]. A significant protective effect than have allowed a prevention of the diarrhea in the neonatal goat kids with the cyclodextrins after oral inoculation of oocytes of C. parvum model [Castro-Hermida *et al.*, 2004]. The cyclodextrins improved the therapy of the disease moreover they acted in prevention and they shown a protective effect at 62% in the chip. The similar improved of the therapeutic index have been shown in other parasite too. The therapeutic equivalence of a beta-cyclodextrin-artemisinin complex have shown similar efficacy than the Artemisinin250 in 100 *Plasmodium falciparum* malaria patients. The beta-cyclodextrin-artemisinin complex at a dose of 150 mg artemisinin was therapeutically equivalent and has comparable plasma artemisinin concentrations than 250 mg Artemisinin250 [Wong *et al.*, 2003]. Nevertheless, their used is limited to the oral application most of the time.

Cyclodextrins are the most widely used to increase the solubility and bioavailability of the small drug but the use for parenteral route is limited.

Two commercial, lipid-associated formulations of amphotericin B are highly effective against visceral leishmaniasis, and better tolerated than conventional preparations. Of the four globally approved commercial formulations of the parenteral administered antibiotic, two are lipid complexes: Abelcet and Amphocil. Abelcet consists of amphotericin B with dimyristoyl phosphatidylcholine and dimyristoyl phosphatidylglycerol. The resulting interdigited rubbon complexes have mean particle diameters of a few microns [Janoff *et al.*, 1993]. The therapeutic index of Abelcet compares favorably with that of Fungizone. Abelcet is quickly removed from circulation by cells of the mononuclear phagocyte system, e.g. by Kupffer cells. A total dose of 10 to 15 mg/kg of Abelcet delivered over 5-10 days cured 90 to 100 per cent of patients [Sundar *et al.*, 1996].

Amphocil is composed of amphotericin B and cholesterol sulfate in equimolar complexes. The discoid particle structure is 4 nm thick, with a mean diameter of 150 nm [Guo, 2001]. Amphocil exhibits an antifungal efficacy similar to Fungizone, and is at the same time less cytotoxic and hemolytic. In addition, amphotericin B administered as Amphocil exhibits a reduced tendency to bind to plasma lipoproteins [Gates *et al.*, 1993]. Significant side effects and signs of nephrotoxicity appear only when daily dosages are $\geq$2.0 mg/kg/day (Fungizone: 0.5–0.75 mg/kg/day). Amphocil followed the same administration scheme as AmBisome, but the dose used showed improved body distribution that resulted in greater *in vivo* efficacy [Yardley *et al.*, 2000]. Nevertheless, the Amphocil *in vivo* instability in some cases increased the level of side effects and intolerance during clinical trials.

These formulations are also used for other parasitic infections that are vulnerable to amphotericin B. These include *Trypanosoma*, *Toxoplasma cryptococcus*, and *echinochocus*. Abelcet increased host survival rates in *Trypanosoma cruzi* mice models, but did not suppress infection in all animals [Yardley *et al.*, 1999]. However, clinical studies have not been undertaken yet.

Other nanodisc-lipid complexes of amphotericin B, Amphodisc, have demonstrated safety and efficacy results similar to those demonstrated by AmBisome [Larabi *et al.*, 2003; Larabi *et al.*, 2004]. Amphotericin B-loaded cochleates complexes were tested for their potential to improve the bioavailability of orally administered drugs in *Leishmania donovani*-infected mice. Amphotericin B loaded-cochleates performed similarly to AmBisome in both oral and intravenously administered forms. However, low costs and ease of use made the oral route the preferred route of administration [Santangelo *et al.*, 2000].

Similar skin-lesion clearance and safety results were obtained when infants were treated with a topical colloidal solution of amphotericin B for 3 weeks. No signs of recurrence were visible even 3 months after cessation of treatment. Amphocil produced the most favorable results in the study, surpassing even Fungizone's performance in mice models [Frankenburg *et al.*, 1998]. These were the first experiments to demonstrate that topical administration of amphotericin B as a complex, either with cholesteryl sulfate or phospholipids, and in the presence of ethanol, could penetrate skin then destroy vulnerable organisms while utilizing low levels of drug concentration. Nevertheless, another clinical trial did not show any improvement of local therapy when Amphocil was used, and amphotericin B could not be detected in plasma after topical application [Vardy *et al.*, 1999]. Differences in ethanol preparation and the interaction of the excipients in Amphocil may explain these contradictory results.

Another therapeutic option is to deliver the drug in aerosol form via the pulmonary route. Beneficial results against fungal infection were obtained in patients after nebulization of Abelcet [Corcoran et al., 2006]. Different other formulations of amphotericin B were clinically tested in pulmonary fungal infections. Most of these clinical studies utilized commonly available commercial lipid formulations of amphotericin B

(AmBisome, Abelcet, Amphocil and Fungizone). The success of amphotericin B in human alveolar echinococcosis after local and/or systemic administration encourages the use of different amphotericin B formulations in large animals model or in clinical trials to confirm their efficacy [Reuter *et al.*, 2003].

3.3. Solid Nanoparticles

Polymer nanoparticles allow a higher level of encapsulation, of different physicochemical properties drugs, than the lipid system. A nanoparticle can reduce the toxicity of effective drugs; modify the bioavailability of therapy in peripheral tissues, which are often a reservoir for parasites in the late stages.

Different poly(lactic acid) nanoparticles and nanocapsules of antiparasitic drugs show improved therapeutic results as compared to free drug forms: primaquine against *Leishmania* [Rodrigues *et al.*, 1994], halofantrine against malaria [Mosqueira *et al.*, 2004], pentamidine against *Leishmania* [Durand *et al.*, 1997]. Different nanoparticles were studied *in vitro* with almost all antiparasite drugs, including poly-epsilon-caprolactone nanospheres [Espuelas *et al.*, 2002] and PLGA nanospheres [Venier-Julienne *et al.*, 1995]. Polymeric nanospheres with doxorubicin Liance *et al.*, 1993] or albendazole [Rodrigues *et al.*, 1995] were developed which demonstrated improved therapy against *Echinococcus multilocularis*. Atovaquone was also encapsulated within a nanocapsule, and showed an efficacy twice that of the free drug form. Liver parasite loads were reduced by 71% when atovaquone was contained within a nanocapsule, versus 34% in the case of the free form [Cauchetier *et al.*, 2003]. Particle size, and the particle composition, led to improved distribution, efficacy, and drug load level for different drugs [Sarkar *et al.*, 2002; Lala *et al.*, 2004]. The optimum formulation was the vesicle in the case of harmine, and nanoparticles in the case of quercetin. Every study revealed a linear correlation between nanoparticle size and efficacy.

Nanospheres made from natural hydrophilic polymers have demonstrated superior drug-loading capacity, biocompatibility, and possibly less opsonization by the reticuloendothelial system (RES) through an aqueous stearic barrier. When the activity of poly(lactic acid) nanoparticles of triterpene against *leishmania donovani* was evaluated, there was an improvement in the efficacy and safety of nanoparticles after subcutaneous injection compared to emulsions. These nanoparticles have shown improved tissue distribution as compared to emulsions, all without inducing the hepatotoxiccity and nephrotoxicity that the emulsion formulations or free drug forms induce [Lala *et al.*, 2004]. Enhancement efficacy has led to colocalization between the infection and the drug incorporated with these carriers [Gaspar *et al.*, 1992; Mosqueira *et al.*, 2004; Rodrigues *et al.*, 1994].

Researchers experimented with various polyesters and found similar results regarding pharmacokinetics and body distribution, and they correlated the colocalization of particles and infectious sites with increased efficacy. Polymethacrylate nanoparticles used to encapsulate pentamidine for *in vivo* testing in experimental *leishmania* infected mice models have shown the same correlation [Fusai *et al.*, 1994]. In parenteral administration, the composition of the specific polymer, one chooses, will be more limited by the drug encapsulation and release. A few were tested *in vivo*, but none in preclinical studies.

The specific surfactant and surface modified polymer nanoparticle improve their pharmacokinetics. Stealth particles demonstrate improved efficacy when compared to itraconazole and ketoconazole encapsulated nanospheres, both of which fare poorly in *Trypanosoma cruzi* infected mice [Molina *et al.*, 2001]. In Brazil, after spending years developing therapeutic nanoparticles for parasitic disease, Rodrigues teams have also developed a 'stealth' nanoparticle composed of polyethylene-glycol-polylactide nanoparticle of bis-triazole DO870 for use against Chagas disease. These nanoparticles have increased the efficacious properties of DO870 in mice models. The result is a cure rate that exceeds the therapeutic reference (benznidazole, 47 and 60% respectively).

Studies of pegylated surface poly(lactic acid) nanocapsules of halothantrine reveal improved therapeutic results in *Plasmodium bergheri* infected mice. The stealth nanocapsules remain in circulation longer. One postulates that stealth nanocapsules will enable doctors to limit the dosage of intravenously administered drugs in severe Malaria cases. These molecules are more efficient in peripheral organs, in the muscle and central nervous system, and on the disseminated localization of the parasite [Mosqueira *et al.*, 2004].

In same time, during the 1990s, many researchers were curious about the immunostimulating effect of nanoparticles *in vitro* and *in vivo*. This effect has clearly been demonstrated *in vitro*. Empty polymeric ethylcyanoacrylate nanoparticles were developed to encapsulate nifurtimox, and they considerably increased the anti-trypanocidal activity in mice [Gonzalez-Martin *et al.*, 1998]. Another study done by Karajgi [Karajgi *et al.*, 1993] has demonstrated a synergy between the nanoparticle and drug *in vivo* therapy against infectious leishmaniasis in rats. These studies show that the encapsulation acts in efficacious synergy with the drug. Different researchers have studied the interaction between nanoparticles and immune cells, in particular with macrophages. Increased secretions of cytotoxic agents like nitric oxide, or the respiratory burst and other radical molecules by the empty nanoparticles, can explain the improved *in vitro* efficacy Venier-Julienne *et al.*, 1995]. However, it does not seem that even so remarkable a burst can totally explain the increased efficacy of these particle systems; the interaction between the particle and the immune system is far more complex.

Orally administered nanocapsules of atovaquone have demonstrated enhanced cure rates in both acute and chronic murine toxoplasmosis (*gondii*), with significantly decreased parasite loads in the brain [Sordet *et al.*, 1998]. The incorporation of arjunglucoside in copolymer N-isopropylacrylamide and N-vinyl pyrrolidone nanogel particles (90 nm) improved therapy in experimental leishmania hamster models [Tyagi *et al.*, 2005]. Nevertheless, similar *in vivo* efficacy was achieved when the drug was encapsulated in 250 nm polyester nanoparticles, but there was reduced hepatotoxiccity and nephrotoxicity [Tyagi *et al.*, 2005].

Another type of nanoparticles, the dendrimers, is a specific nanoparticle that was discovered by Tomalia's team during 1980s. Dendrimers posses multifunction capabilities and can be easily synthesized for topical, oral, or parenteral administration. Some nanoparticles are on the verge being clinically studied for gene therapy, but there is currently no published data about their use in parasitic disease. Nanoscale polyamidoamine dendrimers of appropriate size and charge can be transported across the gastrointestinal epithelial cells with little to no toxicity. The influence of variables such as size, geometry, charge, and drug loading on the transport of polyamidoamine dendrimers across epithelial and endothelial barriers is currently under investigation.

The polymers most often used are polyesters due to their biocompatibility and the fact that they are totally biodegradable. They are also the most well known kinetic of degradation in different mediums and interaction with the blood compound. However, other polymers have specific properties of bioadhesion, larger surface exchange, specific body distribution, hydrophobic drug encapsulation, and stealth. Polymer selection depends on the planned administration route, the target, and the properties of the loaded drug. The most inconvenient of this polymer system resided in the fast clearance by the mononuclear phagocyte system and its reaction with the complement elements that limited the repeated dose of parenteral administration. The process of fabrication still remains more complex than the other formulations; with high level of precipitation or clotting occurs. The advantages of using these particles are site-specific targeting and controlled release of incorporated drugs. However, the cytotoxicity of polymers before and after internalization in cells presents a crucial complication. Further, large-scale production of polymeric nanoparticles is problematic. Therefore, this carrier system thus far remains unpursued by the pharmaceutical industry.

3.4. Lipid Nanosuspensions

The two most developed systems are nanosuspensions, where the drug itself is micronized and solid lipid nanoparticles (SLN) [Müller *et al.*, 2001]. During the mid-90s, different research groups focused on alternative nanoparticles made from solid lipids; solid lipid nanoparticles (SLN). SLNs combine the advantages of other innovative carrier systems (e.g. physical stability, protection of incorporated labile drugs from degradation, controlled release, excellent tolerability) while minimizing associated problems. SLN formulations for various application routes (parenteral, oral, dermal, ocular, pulmonary and rectal) have been developed and thoroughly characterized *in vitro* and *in vivo*.

Two modifications of SLN are nano-structured lipid carriers (NLC) and lipid drug conjugate nanoparticles. These carrier systems can overcome the limitations of conventional SLN. They have the advantage of being physiologically well tolerated in the presence of lipid or fatty acids like the liposome or emulsions respectively, and are very stable as nanoparticles. In addition, they are simple and inexpensive to produce.

Amphotericin B has also been internalized in SLNs that are administered oral and parenteral route. The results have been increase efficacy against Leishmania donovani, increased bioavailability (depending on production parameters), and an average particle size of 100-800 nm. After being formulated as a nanosuspension, Amphotericin B was administered orally and intravenously to L. donovani infected mice. Unlike simple micronized amphotericin B, these nanosuspensions showed appreciable gastrointestinal uptake, and reduced parasite load in the livers of untreated control subjects to 30% for intravenous administration, and 45% for oral administration [Kayser *et al.*, 2003].

When administered intravenously, antileishmanial efficacy of Amphotericin B nanosuspensions resembles that of AmBisome. Amphotericin B nanosuspensions have experienced uptake by phagocyte cell. Electron micrographs have shown parasites and nanoparticles in close vicinity within the same compartment in host cells [Kayser *et al.*, 2001]. Nanoparticles posses clear advantages over liposomes, such as longer shelf lives. Amphotericin B-SLN has improved shelf life at room temperature as compared to other amphotericin B formulations. Oral bioavailability is improved, and production costs are low in comparison. Other amphotericin B SLNs developed near the end of the 1990s have shown excellent efficacy on fungal infections, but few against infectious parasites [Otsubo *et al.*, 1999]. Selective alteration of the physicochemical properties of the particle surface, e.g. by pegylation, might further optimize the pharmacological profile of amphotericin B nanoparticles through a gap of variable width.

Investigations involving mucoadhesive or bioadhesive particles; including hydrogel and polymers, carbopols, polyacrylic acids and chitosan, have been undertaken. Paramomycine or atovaquone encapsulated in mucoadhesive hydrogel adheres to the gastrointestinal wall [Kayser *et al.*, 2001]. Drug-loaded hydrogel can form a film over the entire gastrointestinal tract (GIT), increasing the contact time between pathogen and drug as a function of mucus secretion and biodegradation. A preliminary study has focused on SCID mice, whose is immunodeficiency; they have prevented them from clearing *Cryptosporidium parvum* parasite. In this study, the efficacy of atovaquone or buparvaquone incorporated within a chitosan hydrogel was ten times greater than the efficacy of pure compounds [Kayser *et al.*, 2001]. Although this formulation is simple to produce and can be administered orally, no clinical studies have been undertaken to date. Another type of nanosuspension was developed in early 1990s; ionic amphiphile Biovector formulations. They are composed of the lipid core of DPPG embedded into a polysaccharide film and measure a total size of 100 nm. Different ratios of amphotericin/lipid were tested in *Leishmania donovani* mice models. These nanosuspensions demonstrated *in vivo* parasite inhibition similar to AmBisome [Loiseau *et al.*, 2002].

These nanoparticles show encouraging results. Currently, many related products are being researched, especially orally administered formulations. Their use is currently limited; because of their high stability, the release of the drug is more complex. There is also rapid clearance by the RES after parenteral administration. Vesicles, by comparison, are more flexible, and they allow an exchange with the external medium.

3.5. Liposomes and Niosomes

The most studied vesicules in relation to drug therapy are niosomes and liposomes. Because of their lipid composition, they are well tolerated by host organisms and have acceptable circulation times as compared to polymer systems. There are actually seven commercial liposomes that are used as primary options.

Niosomes are vesicles formed by non-ionic surfactants, and are usually small in size (less than 100 nm in diameter). Niosomes varying in formulation, composition, and charge were all compared. However, the cholesterol content of the vesicle exerted a slight influence on the antiparasitic activity of drug-loaded niosomes, with a similar suppression of liver amastigotes by sodium stibogluconate in murine models involving experimental *Leishmania donovani* [Hunter *et al.*, 1988]. And comparable body distributions to those obtained by free form of the drug were observed after intravenous administration. At equivalent quercetin concentrations, nanocapsulated quercetin most effectively reduced parasite infection loads in the spleen with and overall reduction in hepatotoxcity and renaltoxicity, as compared to free drug forms, or other vesicular drug forms. An inverse relationship efficacy and vesicle size was established [Sarkar *et al.*, 2002]. Another vesicle formulation (sodium stibogluconate noisome) has shown the same efficacy, albeit with reduced distribution in the spleen and bone marrow [Carter *et al.*, 1989]. Efficacy was greater than when free form of the drug were intravenously administered. Parasite loads in the liver, spleen, and bone marrow were diminished; there was greater than 90% parasite suppression in the most of the strains studied [Carter *et al.*, 2001]. Although their efficacy has led to systemic distribution of the drug in *Leishmania* mice models, their failure to reach peripheral sites of infection, such as in bone marrow, in effective concentrations can cause treatment failures and relapse. The association of different surfactants with the sodium stibogluconate has shown better results than the free drug form. Treatment of nonionic surfactant vesicles of sodium stibogluconate results in an improved efficacy *Leishmania spp* infected mice. Noisomes did not demonstrate an exceptional efficacy in drug delivery systems for parenteral administration. Noisome applications are limited, but they are largely used in cutaneous applications in such as cream and they are an alternative local therapy in cutaneous lesions.

The use of liposomes as a new parenteral carrier system is compelling. By the end of the 1970s, many outstanding research teams were publishing the first efficacy results of liposome drug carriers in antimonial treatment for leishmaniasis infected mice [New *et al.*, 1978]. Since then, many other groups have developed a variety of antiparasite drug load-liposomes in the search for a more effective cure. Liposomes can act as carriers for a variety of drugs and can modulate their stability and pharmacokinetic properties, reducing side effects while improving their *in vitro* and *in vivo* activity. Trade products are include AmBisome (1990, Europe, 1997, USA), DaunoXome (1996, Europe and USA) and Doxil (1995, USA, 1996, Europe), Myocet (2000, USA, and Europe) and Pevaryl Lipogel (Europe). In 1993 in UK, AmBisome was the first liposome approved in the world, and in 1997 got FDA approval for use in *Leishmania* infections.

The use of liposomes to deliver antiparasitic drugs meant to treat malaria, trypanosomiasis and Babesia infections is under investigation. So is the use of artemether liposomes as a prolonged release system for beta-artemether, for eradicating recrudescent parasitaemia in Plasmodium chabaudi malaria-infected mice [Chimanuka *et al.*, 2002], and for the encapsulation of chloroquine in liposomes for toxic side effect reduction [Peeters *et al.*, 1989]. Recent in vitro and in vivo studies of the effectiveness of liposomal formulations to treat Trypanosoma brucei ssp. have shown promising results. Specific liposomes consisting of stearylamine and phosphatidylcholine show cytolytic activity against Trypanosoma brucei gambiense in the bloodstream [Tachibana *et al.*, 1988]. Trypanosoma cruzi follows a complex cycle inside a mammalian host, alternating between extra and intracellular stages. Kuboki *et al.* [2006] demonstrated the efficacy of the blank DPPC liposome against African trypanosomes in mice. Other studies have shown the same efficacy of empty liposomes as well [Yoshihara *et al.*, 1987; Tachihara *et al.*, 1988]. Positively charged stearylamine exhibits a higher chemical affinity for cell membranes that leads to improved drug efficacy.

Miltefosine and related alkylphosphocholines can form stable liposomes by themselves if combined with cholesterol and a charged component. Hexadecyl-phoaphatidylcholine, miltefosine (HePC) was formulated in liposomes to reduce side effects, increase its therapeutic index, and to test its use in parenteral administration. New miltefosine liposomal formulations composed of HePC/egg yolk phosphatidylcholine (EPC)/SA 10:10:0.1, 10:10:0.5 and 10:10:1 (molar ratio) have shown *in vitro* activities against *L. donovani* and *T. brucei brucei*, and *in vivo* activity in African trypanosomes [Papagiannaros *et al.*, 2005]. These promising results encouraged researchers to study several alkyl-PCs of varying length and alkyl chain saturation for their effects on

E. histolytica. The alkyl-PC liposomes showed slightly lower activity, but were expected to be well tolerated by patients [Seifert *et al.*, 2001].

Hydrophobic benznidazode is the only antichagasic drug currently approved in Argentina and Brazil. Its extensive binding to plasma proteins causes ubiquitous distribution in almost all types of tissue. This chemical proclivity frustrates efforts to deliver therapeutic amounts of benznidazode to *targeted* cells. Liposomes have been used to potentially increase the selective delivery of this drug. They could be absorbed by infected tissues, which would increase accumulation of benznidazode in the liver [Morilla *et al.*, 2004]. However, to fully exploit the differences between benznidazode and benznidazode liposomes, pharmacokinetic data is still required. If liposomal drugs are administered intravenously, a complex set of phenomena occurs, beginning with the rapid adsorption of plasma proteins. Despite the improvement of the *in vivo* pharmacokinetics of benznidazole encapsulated within multilamellar liposome in rats [Morilla *et al.*, 2005], there was no modification of the parasitaemia in the *Trypanosoma cruzi* mice model. However, encapsulations of nifurtimox, etanidazole, diminazene, miltefosine and allopurinol have shown a reduction of the parasitaemia *in vitro* [Yongsheng *et al.*, 1996; Gonzalez-Martin *et al.*, 1998]. Other studies are equally encouraging; the treatment of *E. histolytica* infections with miltefosine liposomes [Seifert *et al.*, 2001], the treatment of *Toxocara canis* infections in mice with liposome incorporated benzimidazole carbamates [Hrckova *et al.*, 2001], and liposome-encapsulated milbemycins in helminth infections. The third generations of targeted liposomes, chloroquine-containing immuno-liposomes, which use ganglioside antigens, have improved the clearance of *Plasmodium berghei* parasites in red blood cells in rats [Uemura *et al.*, 2006].

Since many arsenic (III) and arsenic (V)-containing compounds are known to demonstrate antiprotozoal properties; researchers are currently exploring the possible antiprotozoal applications of arsonoliposomes in *Trypanosoma* treatment. *In vitro* studies have shown improved therapy against *Trypanosoma brucei* as a result of small arsonoliposomes [Antimisiaris *et al.*, 2003]. Arsonolipids are arsonate-containing lipids, they are analogous to phospholipids, except in this case, arsenic replaces phosphorus. These researchers showed that improved arsonolipid uptake by the parasite membrane significantly influences the efficacy of the arsonoliposome lipid compositions.

Following New's successful studies regarding amphotericin B and leishmania, many amphotericin B liposomes were developed, yet only three progressed to clinical trials. Those however were not well tolerated [Lopez-Berestein *et al.*, 1989; Collette *et al.*, 1991]. A fourth generation, pegylated, targeted liposome of amphotericin was developed and tested in fungal infections. These amphotericin B liposomal formulations reduced toxicity and lengthened serum half-life values [Otsubo *et al.*, 1999; Moribe *et al.*, 2004]. This pegylation induced a hydrodynamic layer on the particle surface, which shielded the liposomes from biological membranes. As a consequence, these particles circulate for longer periods in blood and are experience slower uptake by the mononuclear phagocyte system. Thus, liposomal preparations of amphotericin B, such as AmBisome, are significantly superior to Amphotericin B emulsions or colloidal formulations in terms of bioavailability and side effects. These advantages outweigh the high costs of AmBisome. AmBisome is a unilamellar liposome with a mean diameter of 60–70 nm. The pharmacokinetic profile of AmBisome is significantly superior to that of Fungizone or Abelcet. Effective plasma concentrations are improved, and plasma clearance is therefore significantly reduced. Following parenteral administration, most Amphotericin B is retained in the liver and spleen. Here, liposomes accumulate in cells of the mononuclear phagocyte system via phagocytes. The last ten years have witnessed an increase the use of AmBisome to cure visceral leishmaniasis. Of the three lipid formulations, AmBisome is the best tolerated. AmBisome (2.5mg/kg/day for 20 days in 10 patients) displayed efficacy in the treatment of persistent post-kala-azar dermal leishmaniasis, where 2-4 month regimens of sodium stibogluconate failed [Musa *et al.*, 2005]. A recent study has shown that not only is AmBisome a more effective therapy than sodium stibogluconate having cured disease in one month with no relapse in 12 months, but is also 45% less expensive than sodium stibogluconate. Single administrations of 20 mg/kg to combat *Leishmania donovani* in East Africa also proved successful [Sundar *et al.*, 2000]. Effective single dose treatment makes it possible to treat a large number of patients in a very short time and reduce the cost of therapy and shortened hospital stay. Use of AmBisome in *Trypanosoma cruzi* infected mice increased the survival rates of the mice and suppressed the infection. These findings open a possibility for AmBisome to be used in clinical trials [Yardley *et al.*, 1999].

The liposome stills the more successful nanoparticle and has lot of preparation in the market use in much case in first chose. This liposome shows an increase in the therapeutics index, and is commonly used to treat leishmaniasis, despite the cost. Since their approval for use in fungal infections, the lipid formulations of amphotericin B were successfully used in treating different *Leishmania* infections before and after the approval of melfosine and paramomycine. The formulation scheme will revert to the public domain in few years, and hopefully the price of new generic formulations will render them more affordable.

4. Discussion and Conclusion

Different studies have shown a rate of improved drug efficacy that suggests a synergy between drug and carrier. Different parameters are involved, from the molecular though the carrier characterization, to the *in vivo* interaction: the size, zeta potential, composition, drug aggregation state [Mullen *et al.*, 1997]; polymorphism and ratio to the blood compound interaction [Wasan *et al.*, 1993; Wasan *et al.*, 1994]; tissue distribution, and parasites co-localization, slow release system from the carrier or mononuclear phagocyte system tissues [Venier-Julienne *et al.*, 1995] and immune system reaction. However, no one can strongly conclude and the complex mechanism associate all this property would be certainly involved.

Different nanoparticles have been developed during the past two decades evolving from passive nanoparticles to microspheres to "stealth" systems. These second generation nanoparticles allowed prolonged circulation of the drug, modified its profile of distribution, and allowed better targeting at infection sites. Like the polymer microparticle, this nanoparticle has the advantage of a high level of drug encapsulation as compared to other carrier systems, increase surface, control drug release to infection sites, and better distribution and circulation time in peripheral tissues. Carrier considerations include non-toxicity (acute and chronic), sufficient drug loading capacity, possibility of drug targeting, controlled release characteristics, chemical and physical storage stability (for both drug and carrier) and the economic feasibility to affordably scaling up production. Nanoparticle carriers have generated enormous interest because they can potentially fulfill all of these requirements.

A nanoparticle must be strategically adapted to a particular drug's properties, the pathology of the disease in question, and to the life-cycle of the parasite involved. We cannot simply screen carriers in the laboratory and fall in love with those who give the best results. One should keep in mind that no *one* delivery system per se meets all that is required to solve all the general problems. Optimal formulations must to be chosen carefully for each drug, in accordance with the features of the nanocarriers. The ultimate goal aim is to achieve the desired drug release profiles *in vivo,* all the while minimizing undesired side effects and economic cost.

The fact that the lipid formulation of commercial amphotericin B is still mainly used in wealthy nations challenges researchers to develop cost-effective, accessible and affordable control therapy. One hopes that in few years these formulations will be available in low cost generic forms that will be accessible to nations such as India and Brazil.

Nanoparticle therapy has shown them real possibility and success with the AmBisome the *Leishmania* infectious, than still the more effective therapy. Many avenues of potential remain unexplored despite the present therapy like the pulmonary administration or the nanoparticle topical applications with the nanoparticles already commercialized. Certainly, in few decades, they will show the apparitions of effective and robust vaccines. Unfortunately, the death toll continues to spiral upwards. We must utilize every possible therapy and prevention mechanism that exists against parasites in all of their chosen hosts. Only then can we hope to bring to a halt the catastrophic resurgence of parasitic diseases that now are the leading cause of death worldwide.

References

Antimisiaris, S. G., Ioannou, P. V. and Loiseau, P. M. (2003). In-vitro antileishmanial and trypanocidal activities of arsonoliposomes and preliminary in-vivo distribution in BALB/c mice. *J. Pharm. Pharmacol.*, 55, pp. 647-652.

Beyer, J., Barzen, G., Risse, G., Weyer, C., Miksits, K., Dullenkopf, K., Huhn, D. and Siegert, W. (1993). Aerosol amphotericin B for prevention of invasive pulmonary aspergillosis. *Antimicrob. Agents Chemother.*, 37, pp. 1367-1369.

Brime, B., Frutos, P., Bringas, P., Nieto, A., Ballesteros, M. P. and Frutos, G. (2003). Comparative pharmacokinetics and safety of a novel lyophilized amphotericin B lecithin-based oil-water microemulsion and amphotericin B deoxycholate in animal models. *J. Antimicrob. Chemother.*, 52, pp. 103-109.

Carter, K. C., Dolan, T. F., Alexander, J., Baillie, A. J. and McColgan, C. (1989). Visceral leishmaniasis: drug carrier system characteristics and the ability to clear parasites from the liver, spleen and bone marrow in Leishmania donovani infected BALB/c mice. *J. Pharm. Pharmacol.*, 41, pp. 87-91.

Carter, K. C., Mullen, A. B., Sundar, S. and Kenney, R. T. (2001). Efficacies of vesicular and free sodium stibogluconate formulations against clinical isolates of Leishmania donovani. *Antimicrob. Agents Chemother.*, 45, pp. 3555-3559.

Castro-Hermida, J. A. and Ares-Mazas, M. E. (2003). In vitro and in vivo efficacy of alpha-cyclodextrin for treatment of experimental cryptosporidiosis. *Vet. Parasitol.*, 114, pp. 237-245.

Castro-Hermida, J. A., Pors, I., Otero-Espinar, F., Luzardo-Alvarez, A., Ares-Mazas, E. and Chartier, C. (2004). Efficacy of alpha-cyclodextrin against experimental cryptosporidiosis in neonatal goats. *Vet. Parasitol.*, 120, pp. 35-41.

Castro-Hermida, J. A., Quilez-Cinca, J., Lopez-Bernad, F., Sanchez-Acedo, C., Freire-Santos, F. and Ares-Mazas, E. (2001). Treatment with beta-cyclodextrin of natural Cryptosporidium parvum infections in lambs under field conditions. *Int. J. Parasitol.*, 31, pp. 1134-1137.

Cauchetier, E., Paul, M., Rivollet, D., Fessi, H., Astier, A. and Deniau, M. (2003). Therapeutic evaluation of free and nanocapsule-encapsulated atovaquone in the treatment of murine visceral leishmaniasis. *Ann. Trop. Med. Parasitol.*, 97, pp. 259-268.

Chimanuka, B., Gabriels, M., Detaevernier, M. R. and Plaizier-Vercammen, J. A. (2002). Preparation of beta-artemether liposomes, their HPLC-UV evaluation and relevance for clearing recrudescent parasitaemia in Plasmodium chabaudi malaria-infected mice. *J. Pharm. Biomed. Anal.*, 28, pp. 13-22.

Collette, N., Van der Auwera, P., Meunier, F., Lambert, C., Sculier, J. P. and Coune, A. (1991). Tissue distribution and bioactivity of amphotericin B administered in liposomes to cancer patients. *J. Antimicrob. Chemother.*, 27, pp. 535-548.

Corcoran, T. E., Venkataramanan, R., Mihelc, K. M., Marcinkowski, A. L., Ou, J., McCook, B. M., Weber, L., Carey, M. E., Paterson, D. L., Pilewski, J. M., McCurry, K. R. and Husain, S. (2006). Aerosol deposition of lipid complex amphotericin-B (Abelcet) in lung transplant recipients. *Am. J. Transplant.*, 6, pp. 2765-2773.

Cortadellas, O. (2003). Initial and long-term efficacy of a lipid emulsion of amphotericin B desoxycholate in the management of canine leishmaniasis. *J. Vet. Intern. Med.*, 17, pp. 808-812.

Dangi, J. S., Vyas, S. P. and Dixit, V. K. (1998). Effect of various lipid-bile salt mixed micelles on the intestinal absorption of amphotericin-B in rat. *Drug. Dev. Ind. Pharm.*, 24, pp. 631-635.

Demicheli, C., Ochoa, R., da Silva, J. B., Falcao, C. A., Rossi-Bergmann, B., de Melo, A. L., Sinisterra, R. D. and Frezard, F. (2004). Oral delivery of meglumine antimoniate-beta-cyclodextrin complex for treatment of leishmaniasis. *Antimicrob. Agents Chemother.*, 48, pp. 100-103.

Durand, R., Paul, M., Rivollct, D., Houin, R., Astier, A. and Deniau, M. (1997). Activity of pentamidine-loaded methacrylate nanoparticles against Leishmania infantum in a mouse model. *Int. J. Parasitol.*, 27, pp. 1361-1367.

Espuelas, M. S., Legrand, P., Loiseau, P. M., Bories, C., Barratt, G. and Irache, J. M. (2002). In vitro antileishmanial activity of amphotericin B loaded in poly(epsilon-caprolactone) nanospheres. *J. Drug Target.*, 10, pp. 593-599.

Espuelas, S., Legrand, P., Loiseau, P. M., Bories, C., Barratt, G. and Irache, J. M. (2000). In vitro reversion of amphotericin B resistance in Leishmania donovani by poloxamer 188. *Antimicrob. Agents Chemother.*, 44, pp. 2190-2192.

Frankenburg, S., Glick, D., Klaus, S. and Barenholz, Y. (1998). Efficacious topical treatment for murine cutaneous leishmaniasis with ethanolic formulations of amphotericin B. *Antimicrob. Agents Chemother.*, 42, pp. 3092-3096.

Fusai, T., Deniau, M., Durand, R., Bories, C., Paul, M., Rivollet, D., Astier, A. and Houin, R. (1994). Action of pentamidine-bound nanoparticles against Leishmania on an in vivo model. *Parasite*, 1, pp. 319-324.

Gaboriau, F., Cheron, M., Petit, C. and Bolard, J. (1997). Heat-induced superaggregation of amphotericin B reduces its in vitro toxicity: a new way to improve its therapeutic index. *Antimicrob. Agents Chemother.*, 41, pp. 2345-2351.

Gaspar, R., Opperdoes, F. R., Preat, V. and Roland, M. (1992). Drug targeting with polyalkylcyanoacrylate nanoparticles: in vitro activity of primaquine-loaded nanoparticles against intracellular Leishmania donovani. *Ann. Trop. Med. Parasitol.*, 86, pp. 41-49.

Gates, C. and Pinney, R. J. (1993). Amphotericin B and its delivery by liposomal and lipid formulations. *J. Clin. Pharm. Ther.*, 18, pp. 147-153.

Gonzalez-Martin, G., Merino, I., Rodriguez-Cabezas, M. N., Torres, M., Nunez, R. and Osuna, A. (1998). Characterization and trypanocidal activity of nifurtimox-containing and empty nanoparticles of polyethylcyanoacrylates. *J. Pharm. Pharmacol.*, 50, pp. 29-35.

Grogl, M., Schuster, B. G., Ellis, W. Y. and Berman, J. D. (1999). Successful topical treatment of murine cutaneous leishmaniasis with a combination of paromomycin (Aminosidine) and gentamicin. *J. Parasitol.*, 85, pp. 354-359.

Guo, L. S. (2001). Amphotericin B colloidal dispersion: an improved antifungal therapy. *Adv. Drug Deliv. Rev.*, 47, pp. 149-163.

Hrckova, G. and Velebny, S. (2001). Treatment of Toxocara canis infections in mice with liposome-incorporated benzimidazole carbamates and immunomodulator glucan. *J. Helminthol.*, 75, pp. 141-146.

Hunter, C. A., Dolan, T. F., Coombs, G. H. and Baillie, A. J. (1988). Vesicular systems (niosomes and liposomes) for delivery of sodium stibogluconate in experimental murine visceral leishmaniasis. *J. Pharm. Pharmacol.*, 40, pp. 161-165.

Janoff, A. S., Perkins, W. R., Saletan, S. L. and Swenson, C. E. (1993). Amphotericin B lipid complex (ABLC™): A Molecular rationale for the attenuation of AmB related toxicities. *J. Liposome Res.*, 3, pp. 451-471.

Karajgi, J., Jain, N. K. and Vyas, S. P. (1993). Passive vectoring of a colloidal carrier system for sodium stibogluconate: preparation, characterization and performance evaluation. *J. Drug Target.*, 1, pp. 197-206.

Kayser, O. (2001). A new approach for targeting to Cryptosporidium parvum using mucoadhesive nanosuspensions: research and applications. *Int. J. Pharm.*, 214, pp. 83-85.

Kayser, O., Olbrich, C., Yardley, V., Kiderlen, A. F. and Croft, S. L. (2003). Formulation of amphotericin B as nanosuspension for oral administration. *Int. J. Pharm.*, 254, pp. 73-75.

Kuboki, N., Yokoyama, N., Kojima, N., Sakurai, T., Inoue, N. and Sugimoto, C. (2006). Efficacy of dipalmitoylphosphatidylcholine liposome against African trypanosomes. *J. Parasitol.*, 92, pp. 389-393.

Lala, S., Pramanick, S., Mukhopadhyay, S., Bandyopadhyay, S. and Basu, M. K. (2004). Harmine: evaluation of its antileishmanial properties in various vesicular delivery systems. *J. Drug. Target.*, 12, pp. 165-175.

Lambros, M. P., Bourne, D. W., Abbas, S. A. and Johnson, D. L. (1997). Disposition of aerosolized liposomal amphotericin B. *J. Pharm. Sci.*, 86, pp. 1066-1069.

Larabi, M., Pages, N., Pons, F., Appel, M., Gulik, A., Schlatter, J., Bouvet, S. and Barratt, G. (2004). Study of the toxicity of a new lipid complex formulation of amphotericin B. *J. Antimicrob. Chemother.*, 53, pp. 81-88.

Larabi, M., Yardley, V., Loiseau, P. M., Appel, M., Legrand, P., Gulik, A., Bories, C., Croft, S. L. and Barratt, G. (2003). Toxicity and antileishmanial activity of a new stable lipid suspension of amphotericin B. *Antimicrob. Agents Chemother.*, 47, pp. 3774-3779.

Liance, M., Nemati, F., Bories, C. and Couvreur, P. (1993). Experience with doxorubicin-bound polyisohexylcyanoacrylate nanoparticles on murine alveolar echinococcosis of the liver. *Int. J. Parasitol.*, 23, pp. 427-429.

Loiseau, P. M., Imbertie, L., Bories, C., Betbeder, D. and De Miguel, I. (2002). Design and antileishmanial activity of amphotericin B-loaded stable ionic amphiphile biovector formulations. *Antimicrob. Agents Chemother.*, 46, pp. 1597-1601.

Lopez-Berestein, G., Bodey, G. P., Fainstein, V., Keating, M., Frankel, L. S., Zeluff, B., Gentry, L. and Mehta, K. (1989). Treatment of systemic fungal infections with liposomal amphotericin B. *Arch. Intern. Med.*, 149, pp. 2533-2536.

Maestrelli, F., Gonzalez-Rodriguez, M. L., Rabasco, A. M. and Mura, P. (2005). Preparation and characterisation of liposomes encapsulating ketoprofen-cyclodextrin complexes for transdermal drug delivery. *Int. J. Pharm.*, 298, pp. 55-67.

Maestrelli, F., Mura, P. and Alonso, M. J. (2004). Formulation and characterization of triclosan sub-micron emulsions and nanocapsules. *J. Microencapsul.*, 21, pp. 857-864.

Mbela, T. K., Deharo, E., Haemers, A. and Ludwig, A. (1998). Submicron oil-in-water emulsion formulations for mefloquine and halofantrine: effect of electric-charge inducers on antimalarial activity in mice. *J. Pharm. Pharmacol.*, 50, pp. 1221-1225.

Molina, J., Urbina, J., Gref, R., Brener, Z. and Rodrigues Junior, J. M. (2001). Cure of experimental Chagas' disease by the bis-triazole DO870 incorporated into 'stealth' polyethyleneglycol-polylactide nanospheres. *J. Antimicrob. Chemother.,* 47, pp. 101-104.

Moribe, K. and Maruyama, K. (2002). Pharmaceutical design of the liposomal antimicrobial agents for infectious disease. *Curr. Pharm. Des.*, 8, pp. 441-454.

Morilla, M. J., Montanari, J. A., Prieto, M. J., Lopez, M. O., Petray, P. B. and Romero, E. L. (2004). Intravenous liposomal benznidazole as trypanocidal agent: increasing drug delivery to liver is not enough. *Int. J. Pharm.*, 278, pp. 311-318.

Morilla, M. J., Prieto, M. J. and Romero, E. L. (2005). Benznidazole vs benznidazole in multilamellar liposomes: how different they interact with blood components? *Mem. Inst. Oswaldo Cruz*, 100, pp. 213-219.

Mosqueira, V. C., Loiseau, P. M., Bories, C., Legrand, P., Devissaguet, J. P. and Barratt, G. (2004). Efficacy and pharmacokinetics of intravenous nanocapsule formulations of halofantrine in Plasmodium berghei-infected mice. *Antimicrob. Agents Chemother.*, 48, pp. 1222-1228.

Mullen, A. B., Carter, K. C. and Baillie, A. J. (1997). Comparison of the efficacies of various formulations of amphotericin B against murine visceral leishmaniasis. *Antimicrob. Agents Chemother.*, 41, pp. 2089-2092.

Müller, R. H., Jacobs, C. and Kayser, O. (2001). Nanosuspensions as particulate drug formulations in therapy. Rationale for development and what we can expect for the future. *Adv. Drug Deliv. Rev.*, 47, pp. 3-19.

Musa, A. M., Khalil, E. A., Mahgoub, F. A., Hamad, S., Elkadaru, A. M. and El Hassan, A. M. (2005). Efficacy of liposomal amphotericin B (AmBisome) in the treatment of persistent post-kala-azar dermal leishmaniasis (PKDL). *Ann. Trop. Med. Parasitol*, 99, pp. 563-569.

Nan, A., Croft, S. L., Yardley, V. and Ghandehari, H. (2004). Targetable water-soluble polymer-drug conjugates for the treatment of visceral leishmaniasis. *J. Control. Rel.*, 94, pp. 115-127.

New, R. R., Chance, M. L., Thomas, S. C. and Peters, W. (1978). Antileishmanial activity of antimonials entrapped in liposomes, *Nature*, 272, pp. 55-56.

Otsubo, T., Maesaki, S., Hossain, M. A., Yamamoto, Y., Tomono, K., Tashiro, T., Seki, J., Tomii, Y., Sonoke, S. and Kohno, S. (1999). In vitro and in vivo activities of NS-718, a new lipid nanosphere incorporating amphotericin B, against Aspergillus fumigatus. *Antimicrob. Agents Chemother.*, 43, pp. 471-475.

Ozgoztasi, O. and Baydar, I. (1997). A randomized clinical trial of topical paromomycin versus oral ketoconazole for treating cutaneous leishmaniasis in Turkey. *Int. J. Dermatol.*, 36, pp. 61-63.

Papagiannaros, A., Bories, C., Demetzos, C. and Loiseau, P. M. (2005). Antileishmanial and trypanocidal activities of new miltefosine liposomal formulations. *Biomed. Pharmacother.*, 59, pp. 545-550.

Peeters, P. A., Brunink, B. G., Eling, W. M. and Crommelin, D. J. (1989). Therapeutic effect of chloroquine(CQ)-containing immunoliposomes in rats infected with Plasmodium berghei parasitized mouse red blood cells: comparison with combinations of antibodies and CQ or liposomal CQ. *Biochim. Biophys. Acta*, 981, pp. 269-276.

Petit, C., Yardley, V., Gaboriau, F., Bolard, J. and Croft, S. L. (1999). Activity of a heat-induced reformulation of amphotericin B deoxycholate (fungizone) against Leishmania donovani. *Antimicrob. Agents Chemother.*, 43, pp. 390-392.

Reuter, S., Buck, A., Grebe, O., Nussle-Kugele, K., Kern, P. and Manfras, B. J. (2003). Salvage treatment with amphotericin B in progressive human alveolar echinococcosis. *Antimicrob. Agents Chemother.*, 47, pp. 3586-3591.

Rodrigues, J. M., Jr., Bories, C., Emery, I., Fessi, H., Devissaguet, J. P. and Liance, M. (1995). Development of an injectable formulation of albendazole and in vivo evaluation of its efficacy against Echinococcus multilocularis metacestode. *Int. J. Parasitol.*, 25, pp. 1437-1441.

Rodrigues, J. M., Jr., Croft, S. L., Fessi, H., Bories, C. and Devissaguet, J. P. (1994). The activity and ultrastructural localization of primaquine-loaded poly (d,l-lactide) nanoparticles in Leishmania donovani infected mice. *Trop. Med. Parasitol.*, 45, pp. 223-228.

Santangelo, R., Paderu, P., Delmas, G., Chen, Z. W., Mannino, R., Zarif, L. and Perlin, D. S. (2000). Efficacy of oral cochleate-amphotericin B in a mouse model of systemic candidiasis. *Antimicrob. Agents Chemother.*, 44, pp. 2356-2360.

Sarkar, S., Mandal, S., Sinha, J., Mukhopadhyay, S., Das, N. and Basu, M. K. (2002). Quercetin: critical evaluation as an antileishmanial agent in vivo in hamsters using different vesicular delivery modes. *J. Drug Target.*, 10, pp. 573-578.

Seifert, K., Duchene, M., Wernsdorfer, W. H., Kollaritsch, H., Scheiner, O., Wiedermann, G., Hottkowitz, T. and Eibl, H. (2001). Effects of miltefosine and other alkylphosphocholines on human intestinal parasite Entamoeba histolytica. *Antimicrob. Agents Chemother.*, 45, pp. 1505-1510.

Sordet, F., Aumjaud, Y., Fessi, H. and Derouin, F. (1998). Assessment of the activity of atovaquone-loaded nanocapsules in the treatment of acute and chronic murine toxoplasmosis. *Parasite*, 5, pp. 223-229.

Souza, L. C., Maranhao, R. C., Schreier, S. and Campa, A. (1993). In-vitro and in-vivo studies of the decrease of amphotericin B toxicity upon association with a triglyceride-rich emulsion. *J. Antimicrob. Chemother.*, 32, pp. 123-132.

Sundar, S., Gupta, L. B., Rastogi, V., Agrawal, G. and Murray, H. W. (2000). Short-course, cost-effective treatment with amphotericin B-fat emulsion cures visceral leishmaniasis. *Trans. R. Soc. Trop. Med. Hyg.*, 94, pp. 200-204.

Sundar, S. and Murray, H. W. (1996). Cure of antimony-unresponsive Indian visceral leishmaniasis with amphotericin B lipid complex. *J. Infect. Dis.*, 173, pp. 762-765.

Tachibana, H., Yoshihara, E., Kaneda, Y. and Nakae, T. (1988). In vitro lysis of the bloodstream forms of Trypanosoma brucei gambiense by stearylamine-bearing liposomes. *Antimicrob. Agents Chemother.*, 32, pp. 966-970.

Tasset, C., Preat, V. and Roland, M. (1991). The influence of Myrj 59 on the solubility, toxicity and activity of amphotericin B. *J. Pharm. Pharmacol.*, 43, pp. 297-302.

Torbert, B. J., Kramer, B. S. and Klei, T. R. (1982). Efficacy of injectable and oral paste formulations of ivermectin against gastrointestinal parasites in ponies. *Am. J. Vet. Res.*, 43, pp. 1451-1453.

Trouiller, P., Olliaro, P., Torreele, E., Orbinski, J., Laing, R. and Ford, N. (2002). Drug development for neglected diseases: a deficient market and a public-health policy failure. *Lancet*, 359, pp. 2188-2194.

Tyagi, R., Lala, S., Verma, A. K., Nandy, A. K., Mahato, S. B., Maitra, A. and Basu, M. K. (2005). Targeted delivery of arjunglucoside I using surface hydrophilic and hydrophobic nanocarriers to combat experimental leishmaniasis. *J. Drug Target.*, 13, pp. 161-171.

Uemura, A., Watarai, S., Kushi, Y., Kasama, T., Ohnishi, Y. and Kodama, H. (2006). Analysis of neutral glycosphingolipids from Trypanosoma brucei. *Vet. Parasitol.*, 140, pp. 264-272.

Vakil, R. and Kwon, G. S. (2005). PEG-phospholipid micelles for the delivery of amphotericin B. *J. Control. Rel.*, 101, pp. 386-389.

Vardy, D., Barenholz, Y., Cohen, R., Zvulunov, A., Biton, A., Klaus, S. and Frankenburg, S. (1999). Topical amphotericin B for cutaneous leishmaniasis. *Arch. Dermatol.*, 135, pp. 856-857.

Veerareddy, P. R., Vobalaboina, V. and Nahid, A. (2004). Formulation and evaluation of oil-in-water emulsions of piperine in visceral leishmaniasis. *Pharmazie*, 59, pp. 194-197.

Venier-Julienne, M. C., Vouldoukis, I., Monjour, L. and Benoit, J. P. (1995). In vitro study of the anti-leishmanial activity of biodegradable nanoparticles. *J. Drug Target.*, 3, pp. 23-29.

Wasan, K. M., Brazeau, G. A., Keyhani, A., Hayman, A. C. and Lopez-Berestein, G. (1993). Roles of liposome composition and temperature in distribution of amphotericin B in serum lipoproteins. *Antimicrob. Agents Chemother.*, 37, pp. 246-250.

Wasan, K. M. and Lopez-Berestein, G. (1994). The interaction of liposomal amphotericin B and serum lipoproteins within the biological milieu. *J. Drug Target.*, 2, pp. 373-380.

Wong, J. W., Yuen, K. H., Nagappan, S., Shahul, W. S., Ho, S. S., Gan, E. K. and Toh, W. T. (2003). Therapeutic equivalence of a low dose artemisinin formulation in falciparum malaria patients. *J. Pharm. Pharmacol.*, 55, pp. 193-198.

Yardley, V. and Croft, S. L. (1999). In vitro and in vivo activity of amphotericin B-lipid formulations against experimental Trypanosoma cruzi infections. *Am. J. Trop. Med. Hyg.*, 61, pp. 193-197.

Yardley, V. and Croft, S. L. (2000). A comparison of the activities of three amphotericin B lipid formulations against experimental visceral and cutaneous leishmaniasis. *Int. J. Antimicrob. Agents*, 13, pp. 243-248.

Yongsheng, Y., Yongchun, O., Chengmai, R., Yuanguo, C. and Fenqin, Z. (1996). Trypanocidal value of liposomal diminazene in experimental Trypanosoma brucei evansi infection in mice. *Vet. Parasitol.*, 61, pp. 349-352.

Yoshihara, E., Tachibana, H. and Nakae, T. (1987). Trypanocidal activity of the stearylamine-bearing liposome in vitro. *Life Sci.*, 40, pp. 2153-2159.

Chapter 9

NANOCARRIERS IN THE THERAPY OF INFLAMMATORY DISEASE

Alf Lamprecht

Laboratory of Pharmaceutical Engineering,
Faculty of Medicine and Pharmacy, University of Franche-Comté,
Besançon (France)

1. Introduction

The use of nanoparticulate systems for the delivery of therapeutic agents in inflammatory diseases is receiving considerable attention for medical and pharmaceutical applications. This increasing interest results from the fact that these systems can target more or less selectively inflamed tissue, mainly on a cellular level. Among potential cellular targets by drug-loaded nanoparticles, liposomes, and others macrophages are considered because they play a central role in inflammation. The most common and potent drugs used in macrophage-mediated diseases treatment often induce unwanted side effects, when applied as a free form, due to the necessity of high doses to induce a satisfactory effect. This could result in their systemic spreading, a lack of bioavailability at the desired sites, and a short half-life. Therefore, the use of drug-loaded nanocarriers represents a good alternative to avoid, or at least decrease, side effects and increase efficacy. Here, an overview of the usefulness of nanocarriers for different therapeutic approaches in the treatment of inflammatory diseases is given.

In many cases nanocarrier-based therapy targets the immune related response, namely macrophages. Since the inflammatory reaction shows similarities in different disease forms the drug targeting strategies often resemble. Subsequently, the reader will see in this chapter various strategies for targeting those immune related cells and the related benefit in the several inflammatory diseases.

2. Rheumatoid Arthritis

First approaches in targeted arthritis therapy are rather old. In the beginning, anti-inflammatory drug were encapsulated into liposomes and administered locally to the inflammation site [Dingle *et al.*, 1978; da Silva *et al.*, 1979]. First clinical results of this strategy showed that, intra-articular liposomal cortisol palmitate in a dose equivalent to 2 mg of cortisol produced a worthwhile therapeutic response in patients with rheumatoid arthritis.

Further studies on this strategy using other drugs led also to positive results [Williams *et al.*, 1996]. Arthritis was induced in the right knee joint of Lewis rats. The rats were treated with a single intra-articular injection of different methotrexate formulations. Large multilamellar liposomal preparations of methotrexate were more effective than free drug and small unilamellar vesicles in suppressing inflammation. Their differential effects in treating the antigen-induced arthritis model were related to their retention within the joint space.

Very recently proposed was the amelioration of collagen-induced arthritis in rats by 13-nm gold particles [Tsai *et al.*, 2007]. Nanogold was administered intraarticularly to rats with collagen-induced arthritis before the onset of arthritis. Nanogold bound to vascular endothelial growth factor in arthritis, resulting in inhibition of endothelial cell proliferation and migration. The histologic score (of synovial hyperplasia, cartilage erosion, and leukocyte infiltration), microvessel density, macrophage infiltration, and levels of TNFalpha and IL-1beta were significantly reduced in the ankle joints of nanogold-treated rats. Nanogold exerted antiangiogenic activities and subsequently reduced macrophage infiltration and inflammation, which resulted in attenuation of arthritis.

Polymeric nanoparticles were introduced into the arthritis recently [Kim *et al.*, 2002; Horisawa *et al.*, 2002]. Polyester nanoparticles were tested in order to investigate the tolerogenic effect of single administration of nanoparticles entrapping type II collagen on the development of collagen-induced arthritis. After single oral administration of nanoparticles, numerous particles approximately 300 nm in size were detectable in Peyer's patches 14 days after the original feeding. Nanoparticle formulations were also able to suppress arthritis after disease onset. Moreover, nanoparticle treated mice showed a higher level of TGFbeta mRNA expression in Peyer's patches, but a lower level of TNFalpha mRNA expression in draining lymph nodes, compared with the other groups of mice. This approach may hold promise as a new treatment strategy in rheumatoid arthritis, but further in-depth studies were not reported on this therapeutic strategy.

Besides, nanoparticles were proposed as an intra-articular delivery system for betamethasone in an ovalbumin-induced chronic synovitis model in the rabbit [Horisawa *et al.*, 2002]. In the antigen-induced arthritic rabbit, the joint swelling decreased significantly by administering betamethasone-loaded nanospheres during a 21-day period after intra-articular challenge. Betamethasone-loaded nanospheres were phagocytosed by the synovial activated-cells and the cartilage degradation was almost prevented.

A different approach was reported on the basis of chitosan-mediated gene delivery to the rabbit knee joints [Zhang *et al.*, 2006]. This study first sought to confirm that foreign genes can be transferred to articular chondrocytes in primary culture. Next, chitosan-DNA nanoparticles containing IL-1Ra genes were injected directly into the knee joint cavities of osteoarthritis rabbits to clarify the in vivo transfer availability of the vectors. Clear expression of IL-1Ra was detected in the knee joint synovial fluid of the chitosan IL-1Ra-injected group. A significant reduction was also noted in the severity of histologic cartilage lesions in the group that received the chitosan IL-1Ra injection.

These locally administered nanocarriers showed all a significant therapeutic effect and would be promising as a therapeutic tool. However, their major drawback remains the need of a direct injection into the inflamed tissue, which can be delicate.

Another drug delivery approach is the selective accumulation of long-circulating liposomes at the inflammation site after intravenous administration. Several preclinical studies dealt with this strategy.

Liposomal prednisolone proved to be highly effective in the rat adjuvant-induced arthritis model [Metselaar *et al.*, 2003]. A single injection of 10 mg/kg resulted in complete remission of the inflammatory response for almost a week. In contrast, the same dose of unencapsulated prednisolone did not reduce inflammation, and only a slight effect was observed after repeated daily injections. Evidence was found that preferential glucocorticoid delivery to the inflamed joint was the key factor explaining the observed strong therapeutic benefit obtained with the liposomal preparation, while other possible mechanisms, such as splenic accumulation or prolonged release of prednisolone in the circulation, were excluded.

In terms of mechanism of action a more detailed study showed the following [Metselaar *et al.*, 2004]: Mice with collagen type II-induced arthritis treated with 10 mg/kg liposomal prednisolone resulted in a strong and lasting resolution of joint inflammation. 10 mg/kg free prednisolone only became slightly effective after repeated daily injections. Although joint inflammation recurred 1 week after treatment with liposomal prednisolone, knee joint sections prepared at this time indicated that the cartilage damage was still reduced. Localisation of gold labelled liposomes in the inflamed joints was seen in the proximity of blood vessels, in the cellular infiltrate, but mainly in the synovial lining. Unaffected joints did not take up liposomes.

Liposomal formulations of acylated superoxide dismutase and unmodified superoxide dismutase were prepared on the basis of two types of liposomes: conventional liposomes presenting an unmodified external surface and long circulating liposomes [Gaspar *et al.*, 2007]. The 'enzymosomes' are nanocarriers combining the advantages of expressing enzymatic activity in intact form and thus being able to exert therapeutic effect even before liposomes disruption, as well as acting as a sustained release of the enzyme.

In order to further increase the specificity of the drug targeting, dexamethasone-loaded long-circulating liposomes were modified exposing on their surface RGD peptides targeted to alphavbeta3 integrins expressed on angiogenic vascular endothelial cells and subsequently able to bind this cell type at inflammation sites [Koning *et al.*, 2006]. RGD-liposomes were reported to bind and to be taken up by proliferating human VECs in vitro. In vivo, increased targeting to areas of lipopolysaccharide-induced inflammation in rats was observed. Specific association with the blood vessel wall at the site of inflammation was confirmed by intravital microscopy. One single intravenous injection of dexamethasone-loaded RGD-liposomes resulted in a strong and long-lasting antiarthritic effect in rats.

Another recent approach was based on the selective liposomal delivery of siRNA [Khoury *et al.*, 2006]. TNFalpha is among the most prominent cytokines in rheumatoid arthritis and is secreted mainly by macrophages. A direct method for restoring the immunologic balance in arthritis is use of small interfering RNA (siRNA) for silencing the TNFalpha transcript. The aim of this study was to determine the therapeutic effect of systemic administration of TNFalpha siRNA in an experimental model of arthritis, optimizing its delivery using new liposome formulations. *In vivo*, complete cure of arthritis was observed when TNFalpha siRNA was administered weekly, complexed with the liposome and combined with carrier DNA. Inhibition (50-70%) of articular and systemic TNFalpha secretion was detected in the siRNA-injected groups, which correlated with a decrease in the levels of IL-6 and monocyte chemotactic protein 1.

When administering slightly different nanoparticles loaded with betamethasone by intravenous route also distinct therapeutic effect was observed [Higaki *et al.*, 2005]. In adjuvant arthritis rats, a 30% decrease in paw inflammation was obtained in 1 day and maintained for 1 week with a single injection of 100 μg of nano-steroid. X-ray examination 7 days after this treatment showed decreased soft tissue swelling. In contrast, the same dose of free betamethasone after three administrations only moderately reduced the severity of inflammation. In addition, a histological examination 7 days after the treatment showed a significant decrease of the inflammatory cells in the joints. The immune suppressive drug triptolide was entrapped into nanoparticles in order to increase its therapeutic index and to reduce adverse effects [Liu *et al.*, 2005]. The results obtained in experiments indicated that nanoparticles significantly inhibited the adjuvant-induced arthritis, and had preferable anti-inflammatory effect with the long-time administration.

Also dendrimers have been applied for selective drug delivery to the inflammation site in arthritis. In this study indomethacin was linked to poly(amidoamine) dendrimers [Chauhan *et al.*, 2004]. Intravenous administration of the drug-loaded dendrimer in rats showed a two-compartment pharmacokinetic profile. Enhanced effective indomethacin concentrations in the inflamed regions were obtained for the prolonged time period with the indomethacin-loaded dendrimer complex compared to the free drug in arthritic rats indicating its preferred accumulation. The targeting efficiency 2.29 times higher compared to free drug. Moreover, inspite of lymphatic drainage, retention of dendrimers occurs at the inflammatory site.

3. Inflammatory Bowel Disease

Since anti-inflammatory and more recently used immune suppressive drugs are known for their distinct adverse effects, local drug delivery towards the site of inflammation is indispensable in the therapy of inflammatory bowel disease (IBD). The fact that ulcerative colitis and Crohn's disease affect limited areas of the distal intestine underlying distinct interindividual variability of the inflammation site turns drug therapy complicated. Although many efforts have been made in the development of specific drug delivery systems, classical drug delivery systems are still not completely successful. Beside incidences where the therapy fails due to insufficient drug concentrations at the site of action, adverse drug effects have been observed, which act as limiting factors for the respective therapy. These adverse effects are thought to be related to the lack of selective drug release as conventional colon delivery is triggered by factors widely independent from physiological conditions of the inflammation and its location. Consequently, distinct drug loads are delivered unintentionally to areas with non-inflamed tissue during intestinal passage of the drug carrier. While drugs delivered towards the inflamed tissue mitigate the disease, healthy tissue surrounding the site of inflammation risk to absorb the drug, potentially provoking adverse reactions.

Several studies have indicated the strong involvement of macrophages and dendritic cells at the inflammation site of active IBD. A new therapeutic approach is proposed on the basis of this cellular immune response occurring in the inflamed regions, in general, an increased presence of neutrophils, natural killer cells, mast cells, and regulatory T cells [Allison *et al.*, 1988; Seldenrijk *et al.*, 1989]. In consequence, it was hypothesized that particle uptake into those immune-related cells or the disrupted intestinal barrier at ulcerated regions [Stein *et al.*, 1998] could allow the selective accumulation of the particulate carrier system in the desired area (Figure 1).

Subsequently, the size-dependent deposition of microparticles and nanoparticles after oral administration to rats using an experimental model colitis was examined with the aim of the development of a strategy of selective drug delivery [Lamprecht *et al.*, 2001]. In the inflamed tissue, an increased adherence of particles was observed at thicker mucus layers and in the ulcerated regions with a size dependency of the deposition compared with the control group. The ratio of colitis/control deposition increased with smaller particle sizes.

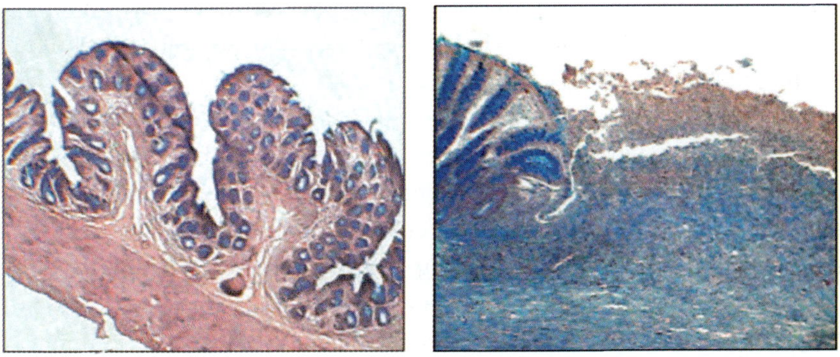

(a) (b)

Fig. 1 Examples for histologic colon sections of healthy (a), untreated TNBS colitis (b), in rats.

In the following studies, therapeutic efficiency of this approach was analysed [Lamprecht *et al.*, 2001; Lamprecht *et al.*, 2005]. Nanoparticles were especially intended for targeted drug delivery to the inflammation site in severe cases of IBD where state-of-the-art delivery devices fail. The drug loaded nanoparticle formulations enabled the drug to accumulate in the inflamed tissue with higher efficiency than when given as solution. Further mechanistic studies showed that tacrolimus loaded nanoparticles allow an enhanced and selective drug penetration into the inflammation site as opposed to surrounding healthy tissue, presumably by protecting the encapsulated drug against influences from efflux systems and mucosal metabolism [Lamprecht *et al.*, 2005]. The relative drug penetration into the inflamed tissue is about 3-fold higher compared with healthy tissue when using nanoparticles as drug carriers.

In comparative study between polyester nanoparticles, similar to those administered above, and pH-sensitive nanoparticles therapeutic efficiency was not significantly different from both therapeutic approaches [Meissner *et al.*, 2006]. Free drug receiving groups (oral/subcutaneous) exhibited increased levels of adverse effects, whereas both nanoparticle types demonstrated their potential to reduce nephrotoxicity. The consequences from these observations are that the involved mechanisms are considered far more complex. Subsequently, it is impossible to estimate the efficiency of this therapeutic approach. Espeically clinical aspects need to be elucidated.

On the basis of cyclodextrines a promising approach was developed for the thrapy of ulcerative colitis [Yano *et al.*, 2001]. Prednisolone-appended cyclodextrins were tested for their therapeutic efficiency after intracolonic administration in experimental colitis in rats. The local anti-inflammatory activity increased in the order of prednisolone alone = prednisolone alpha-cyclodextrin conjugate < prednisolone beta-cyclodextrin complex. As to systemic adverse effect, the prednisolone beta-cyclodextrin and prednisolone alone caused thymolysis at doses of 5-10 mg/kg while the prednisolone alpha-cyclodextrin conjugates showed no clear systemic adverse effect. The low adverse effect of the conjugate may be ascribed to the slow release of prednisolone in the colon, which keeps the local concentration in the colon at a low but constant level.

Recent approaches by different liposomal drug delivery approaches showed also distinct success. Based on adherence to intestinal mucosa, intralumenally administered liposomal formulations of 5-aminosalicylate and 6-mercaptopurine were studied for their potential to enhance local drug delivery to intestinal tissue for the treatment of IBD [Kesisoglou *et al.*, 2005]. Liposomal adherence to intestinal tissue resulted in increased tissue levels for 5-aminosalicylate; however, 6-mercaptopurine local tissue levels were not improved compared to solution drug. While liposomal formulations show potential for local drug delivery to diseased bowel, drug physicochemical properties, absorption, and metabolic profiles dictate tissue-targeting potential.

Differences in liposome's surface charge were also exploited for a targeted drug delivery strategy in experimental colitis [Jubeh *et al.*, 2004; Jubeh *et al.*, 2006]. Superoxide dismutase, 4-amino tempol, and catalase were encapsulated into negatively charged liposomes. The activity of the antioxidants in experimental colitis was tested in rats and compared to the anti-inflammatory activity of the native enzymes and free 4-amino tempol. In all cases, the liposomal preparations of the antioxidants were more effective than the free molecules in the treatment of the experimental colitis, probably due to the attachment of the negatively charged liposomes, and consequently a longer residence time and better uptake of the antioxidants to the inflamed mucosa.

Carnitine transporters have recently been implicated in susceptibility to IBD. Because carnitine is required for beta-oxidation, it was suggested that decreased carnitine transporters, and hence reduced carnitine uptake, could lead to impaired fatty acid oxidation in intestinal epithelial cells, and to cell injury. Treatment with carnitine-loaded liposomes corrected the butyrate metabolic alterations *in vitro* and reduced the severity of colitis in vivo [D'Argenio *et al.*, 2006]. These results suggest that carnitine depletion in colonocytes is associated with the inability of mitochondria to maintain normal butyrate beta-oxidation. It remains to confirm whether this approach may lead to a therapeutic development.

Intravenously administered liposomes showed an increased uptake in the inflamed colonic tissue owing the endothelium fenestration in the inflamed area and were able to visualize colitic lesions [Oyen *et al.*, 1997]. First, liposomes were used to evaluate the extent and severity of abnormalities in IBD. Radiolabeled liposomes were given to animals suffering from a model colitis followed by scintigraphic evaluation. These liposomal formulations possess "adhesion" properties and therefore have attracted attention as drug delivery system for an endothelial delivery of drugs towards the inflammation site. In consequence, liposomes have been developed as drug targeting agents for the treatment of IBD [Awasthi *et al.*, 2002]. Injected poly(ethylene glycocol)-liposomes preferentially accumulated in the inflamed tissue of colitis rats (around 13%), against 0.1% in the normal region of the control group.

It was shown recently that the up-regulation of endothelial cell adhesion molecules can be exploited to selectively target the inflamed endothelium which means a targeting approach from the "backside". Particles made from a biodegradable block copolymer of poly(lactic acid) and poly(ethylene glycocol), to which ligands to these adhesion molecules were conjugated, exhibit specific and augmented adhesion to inflamed endothelium relative to non-inflamed endothelium in vitro and in vivo. Also the specific targeting to vascular cell adhesion molecules-1 in a murine colitis model was demonstrated [Sakhalkar *et al.*, 2003]. The prepared systems proved to significantly enhance particle adhesion to the inflamed endothelium whereas selectivity and ligand efficiency was dependent to the number of particles injected. However, it remains to be proven whether this approach is efficient enough to reach sufficiently high drug levels in the inflammation site and above all, is able to avoid adverse effects.

4. Uveitis

First, Ketorolac entrapped in polymeric micelles was proposed in ocular anti-inflammatory studies [Gupta *et al.*, 2000]. Polymeric micelles made of copolymer of N-isopropylacrylamide, vinyl pyrrolidone and acrylic acid having cross-linkage with N,N'-methylene bis-acrylamide were used as carrier in which up to 30% ketorolac (free acid) was entrapped. In vitro corneal permeation studies through excised rabbit cornea indicated two fold increase in ocular availability with no corneal damage compared to an aqueous suspension containing same amount of drug as in nanoparticles. The formulation showed significant inhibition of lid closure up to 3 h and neutrophil migration up to 5 h compared to the suspension containing non-entrapped drug, which did not show any significant effect.

Others proposed an enhanced ocular anti-inflammatory activity by ibuprofen loaded Eudragit® RS100 nanoparticle suspension after topical administration [Bucolo *et al.*, 2002]. The ibuprofen nanosuspension significantly reduced the primary signs of ocular inflammation as well as significantly reducing the protein level and the number of polymorphonuclear leukocytes in the aqueous humor compared with free ibuprofen. Furthermore, the aqueous humor drug concentration from the group treated with ibuprofen nanoparticles was significantly higher compared to the free drug group.

In a recent study, intraocular injection of tamoxifen-loaded nanoparticles were proposed as a new treatment of experimental autoimmune uveoretinitis [de Kozak *et al.*, 2004]. To increase its bioavailability tamoxifen was incorporated into polyethylene glycol-coated nanoparticles. Some nanoparticles were distributed extracellularly throughout the ocular tissues, others were concentrated in resident ocular cells and in infiltrating macrophages. Whereas the injection of free tamoxifen did not alter the course of autoimmune uveoretinitis, injection of drug loaded nanoparticles performed before the onset of the disease resulted in significant inhibition. Low expression of TNF-alpha, IL-1beta, and RANTES mRNA were noted in eyes of nanoparticle-treated rats. Intravitreal injection of tamoxifen-loaded nanoparticles decreased S-Ag lymphocyte proliferation, IFN-gamma production by inguinal lymph node cells, and specific delayed-type hypersensitivity indicative of a reduced Th1-type response. It increased the anti-S-Ag IgG1 isotype indicating an antibody class switch to Th2 response.

Another study entrapped betamethasone in poly(lactic acid) nanoparticles [Sakai *et al.*, 2006]. The authors developed nanoparticles, which were capable of targeting a specific lesion and gradually releasing the agent at the site over a prolonged time period after a single intravenous administration for local delivery in experimental autoimmune uveoretinitis in rats. Intravenously injected nanoparticles accumulated in the retina and choroid of rats with autoimmune uveoretinitis within 3 hours and remained over the succeeding 7-day-period. Furthermore, systemically administered nanoparticles reduced the clinical scores of rats within 1 day, which were maintained for 2 weeks and decreased the histological scores. In addition, the ocular infiltration of activated T-cells and macrophages were markedly reduced with this treatment. Systemically administered betamethasone loaded nanoparticles inhibited the development of autoimmune uveoretinitis due to the targeting and the sustained release of steroids in situ.

Considering all these different remarkable approaches in the therapy of inflammatory disease, the question arises for the reasons for the low number of clinical trials elucidating the efficiency of these approaches in humans.

References

Allison, M. C., Cornwall, S., Poulter, L. W., Dhillon, A. P. and Pounder, R. E. (1988). Macrophage heterogeneity in normal colonic mucosa and in inflammatory bowel disease. *Gut*, 29, pp. 1531-1538.

Awasthi, V. D., Goins, B., Klipper, R. and Phillips, W. T. (2002). Accumulation of PEG-liposomes in the inflamed colon of rats: potential for therapeutic and diagnostic targeting of inflammatory bowel diseases. *J. Drug Target.*,10, pp. 419-427.

Bourges, J. L., Gautier, S. E., Delie, F., Bejjani, R. A., Jeanny, J. C., Gurny, R., BenEzra, D. and Behar-Cohen, F. F. (2003). Ocular drug delivery targeting the retina and retinal pigment epithelium using polylactide nanoparticles. *Invest. Ophthalmol. Vis. Sci.*, 44, pp. 3562-3569.

Bucolo, C., Maltese, A., Puglisi, G. and Pignatello, R. (2002). Enhanced ocular anti-inflammatory activity of ibuprofen carried by an Eudragit RS100 nanoparticle suspension. *Ophthalmic. Res.*, 34, pp. 319-323.

Chauhan, A. S., Jain, N. K., Diwan, P. V. and Khopade, A. J. (2004). Solubility enhancement of indomethacin with poly(amidoamine) dendrimers and targeting to inflammatory regions of arthritic rats. *J. Drug Target.*, 12, pp. 575-583.

Chellat, F., Merhi, Y., Moreau, A. and Yahia, L. (2005). Therapeutic potential of nanoparticulate systems for macrophage targeting. *Biomaterials*, 26, pp. 7260-7275.

D'Argenio, G., Calvani, M., Casamassimi, A., Petillo, O., Margarucci, S., Rienzo, M., Peluso, I., Calvani, R., Ciccodicola, A., Caporaso, N. and Peluso, G. (2006). Experimental colitis: decreased Octn2 and Atb0+ expression in rat colonocytes induces carnitine depletion that is reversible by carnitine-loaded liposomes. *FASEB J.*, 20, pp. 2544-2546.

de Kozak, Y., Andrieux, K., Villarroya, H., Klein, C., Thillaye-Goldenberg, B., Naud, M. C., Garcia, E. and Couvreur, P. (2004). Intraocular injection of tamoxifen-loaded nanoparticles: a new treatment of experimental autoimmune uveoretinitis. *Eur. J. Immunol.*, 34, pp. 3702-3712.

de Silva, M., Hazleman, B. L., Thomas, D. P. and Wraight, P. (1979). Liposomes in arthritis: a new approach. *Lancet*, 1, pp. 1320-1322.

Dingle, J. T., Gordon, J. L., Hazleman, B. L., Knight, C. G., Page Thomas, D. P., Phillips, N. C., Shaw, I. H., Fildes, F. J., Oliver, J. E., Jones, G., Turner E. H. and Lowe, J. S. (1978). Novel treatment for joint inflammation. *Nature*, 271, pp. 372-373.

Gaspar, M. M., Boerman, O. C., Laverman, P., Corvo, M. L., Storm, G. and Cruz, M. E. (2007). Enzymosomes with surface-exposed superoxide dismutase: in vivo behaviour and therapeutic activity in a model of adjuvant arthritis. *J. Control. Rel.*, 117, pp. 186-195.

Gupta, A. K., Madan, S., Majumdar, D. K. and Maitra, A. (2000). Ketorolac entrapped in polymeric micelles: preparation, characterisation and ocular anti-inflammatory studies. *Int. J. Pharm.*, 209, pp. 1-14.

Harigai, T., Hagiwara, H., Ogawa, Y., Ishizuka, T., Kaneda, S. and Kimura, J. (2007). Prednisolone phosphate-containing TRX-20 liposomes inhibit cytokine and chemokine production in human fibroblast-like synovial cells: a novel approach to rheumatoid arthritis therapy. *J. Pharm. Pharmacol.*, 59, pp. 137-143.

Hattori, Y., Sakaguchi, M. and Maitani, Y. (2006). Folate-linked lipid-based nanoparticles deliver a NFkappaB decoy into activated murine macrophage-like RAW264.7 cells. *Biol. Pharm. Bull.*, 29, pp. 1516-1520.

Higaki, M., Ishihara, T., Izumo, N., Takatsu, M. and Mizushima, Y. (2005). Treatment of experimental arthritis with poly(D, L-lactic/glycolic acid) nanoparticles encapsulating betamethasone sodium phosphate. *Ann. Rheum. Dis.*, 64, pp. 1132-1136.

Horisawa, E., Hirota, T., Kawazoe, S., Yamada, J., Yamamoto, H., Takeuchi, H. and Kawashima, Y. (2002). Prolonged anti-inflammatory action of DL-lactide/glycolide copolymer nanospheres containing betamethasone sodium phosphate for an intra-articular delivery system in antigen-induced arthritic rabbit. *Pharm. Res.*, 19, pp. 403-410.

John, A. E., Lukacs, N. W., Berlin, A. A., Palecanda, A., Bargatze, R. F., Stoolman, L. M. and Nagy, J. O. (2003). Discovery of a potent nanoparticle P-selectin antagonist with anti-inflammatory effects in allergic airway disease. *FASEB J.*, 17, pp. 2296-2298.

Jubeh, T. T., Barenholz, Y. and Rubinstein, A. (2004). Differential adhesion of normal and inflamed rat colonic mucosa by charged liposomes. *Pharm. Res.*, 21, pp. 447-453.

Jubeh, T. T., Nadler-Milbauer, M., Barenholz, Y. and Rubinstein, A. (2006). Local treatment of experimental colitis in the rat by negatively charged liposomes of catalase, TMN and SOD. *J. Drug Target.*,14, pp. 155-163.

Kesisoglou, F., Zhou, S. Y., Niemiec, S., Lee, J. W., Zimmermann, E. M. and Fleisher, D. (2005). Liposomal formulations of inflammatory bowel disease drugs: local versus systemic drug delivery in a rat model. *Pharm. Res.*, 22, pp. 1320-1330.

Khoury, M., Louis-Plence, P., Escriou, V., Noel, D., Largeau, C., Cantos, C., Scherman, D., Jorgensen, C. and Apparailly, F. (2006). Efficient new cationic liposome formulation for systemic delivery of small interfering RNA silencing tumor necrosis factor alpha in experimental arthritis. *Arthritis Rheum.*, 54, pp. 1867-1877.

Kim, W. U., Lee, W. K., Ryoo, J. W., Kim, S. H., Kim, J., Youn, J., Min, S. Y., Bae, E. Y., Hwang, S. Y., Park, S. H., Cho, C. S., Park, J. S. and Kim, H. Y. (2002). Suppression of collagen-induced arthritis by single administration of poly(lactic-co-glycolic acid) nanoparticles entrapping type II collagen: a novel treatment strategy for induction of oral tolerance. *Arthritis Rheum.*, 46, pp. 1109-1120.

Koning, G. A., Schiffelers, R. M., Wauben, M. H., Kok, R. J, Mastrobattista, E., Molema, G., ten Hagen, T. L. and Storm, G. (2006). Targeting of angiogenic endothelial cells at sites of inflammation by dexamethasone phosphate-containing RGD peptide liposomes inhibits experimental arthritis. *Arthritis Rheum.*, 54, pp. 1198-1208.

Lamprecht, A., Schafer, U. and Lehr, C. M. (2001). Size-dependent bioadhesion of micro- and nanoparticulate carriers to the inflamed colonic mucosa. *Pharm. Res.*, 18, pp. 788-793.

Lamprecht, A., Ubrich, N., Yamamoto, H., Schafer, U., Takeuchi, H., Maincent, P., Kawashima, Y. and Lehr C. M. (2001). Biodegradable nanoparticles for targeted drug delivery in treatment of inflammatory bowel disease. *J. Pharmacol. Exp. Ther.*, 299, pp. 775-781.

Lamprecht, A., Yamamoto, H., Takeuchi, H. and Kawashima, Y. (2005). Nanoparticles enhance therapeutic efficiency by selectively increased local drug dose in experimental colitis in rats. *J. Pharmacol. Exp. Ther.*, 315, pp. 196-202.

Liu, M., Dong, J., Yang, Y., Yang, X. and Xu, H. (2005). Anti-inflammatory effects of triptolide loaded poly(D,L-lactic acid) nanoparticles on adjuvant-induced arthritis in rats. *J. Ethnopharmacol.*, 97, pp. 219-25.

Meissner, Y., Pellequer, Y. and Lamprecht, A. (2006). Nanoparticles in inflammatory bowel disease: particle targeting versus pH-sensitive delivery. *Int. J. Pharm.*, 316, pp. 138-143.

Merodio, M., Irache, J. M., Eclancher, F., Mirshahi, M. and Villarroya, H. (2000). Distribution of albumin nanoparticles in animals induced with the experimental allergic encephalomyelitis. *J. Drug Target.*, 8, pp. 289-303.

Merodio, M., Irache, J. M., Valamanesh, F. and Mirshahi, M. (2002). Ocular disposition and tolerance of ganciclovir-loaded albumin nanoparticles after intravitreal injection in rats, *Biomaterials*, 23, pp. 1587-1594.

Metselaar, J. M., van den Berg, W. B., Holthuysen, A. E., Wauben, M. H., Storm, G. and van Lent, P. L. (2004). Liposomal targeting of glucocorticoids to synovial lining cells strongly increases therapeutic benefit in collagen type II arthritis. *Ann. Rheum. Dis.*, 63, pp. 348-353.

Metselaar, J. M., Wauben, M. H., Wagenaar-Hilbers, J. P., Boerman; O. C. and Storm; G. (2003). Complete remission of experimental arthritis by joint targeting of glucocorticoids with long-circulating liposomes. *Arthritis Rheum.*, 48, pp. 2059-2066.

Oyen, W. J., Boerman, O. C., Dams, E. T., Storm, G., van Bloois, L., Koenders, E. B., van Haelst, U. J., van der Meer, J. W. and Corstens, F. H. (1997). Scintigraphic evaluation of experimental colitis in rabbits. *J. Nucl. Med.*, 38, pp. 1596-1600.

Sakai, T., Kohno, H., Ishihara, T., Higaki, M., Saito, S., Matsushima, M., Mizushima, Y. and Kitahara, K. (2006). Treatment of experimental autoimmune uveoretinitis with poly(lactic acid) nanoparticles encapsulating betamethasone phosphate. *Exp. Eye Res.*, 82, pp. 657-663.

Sakhalkar, H. S., Dalal, M. K., Salem, A. K., Ansari, R., Fu, J., Kiani, M. F., Kurjiaka, D. T., Hanes, J., Shakesheff, K. M. and Goetz, D. J. (2003). Leukocyte-inspired biodegradable particles that selectively and avidly adhere to inflamed endothelium in vitro and in vivo. *Proc. Natl. Acad. Sci.*, 100, pp. 15895-15900.

Seldenrijk, C. A., Drexhage, H. A., Meuwissen, S. G., Pals, S. T. and Meijer, C. J. (1989). Dendritic cells and scavenger macrophages in chronic inflammatory bowel disease. *Gut*, 30, pp. 484-491.

Stein, J., Ries, J. and Barrett, K. E. (1998). Disruption of intestinal barrier function associated with experimental colitis: possible role of mast cells. *Am. J. Physiol.*, 274, pp. G203-209.

Tsai, C. Y., Shiau, A. L., Chen, S. Y., Chen, Y. H., Cheng, P. C., Chang, M. Y., Chen, D. H., Chou, C. H., Wang, C. R. and Wu, C. L. (2007). Amelioration of collagen-induced arthritis in rats by nanogold. *Arthritis Rheum.*, 56, pp. 544-554.

Tsurumoto, T., Matsumoto, T., Yonekura, A. and Shindo, H. (2006). Nanobacteria-like particles in human arthritic synovial fluids. *J. Proteome Res.*, 5, pp. 1276-1278.

Williams, A. S., Camilleri, J. P., Goodfellow, R. M. and Williams, B. D. (1996). A single intra-articular injection of liposomally conjugated methotrexate suppresses joint inflammation in rat antigen-induced arthritis. *Br. J. Rheumatol.*, 35, pp. 719-724.

Yano, H., Hirayama, F., Arima, H. and Uekama, K. (2001). Prednisolone-appended alpha-cyclodextrin: alleviation of systemic adverse effect of prednisolone after intracolonic administration in 2,4,6-trinitrobenzenesulfonic acid-induced colitis rats. *J. Pharm. Sci.*, 90, pp. 2103-2112.

Zhang, X., Yu, C., Xushi, Zhang, C., Tang, T. and Dai K. (2006). Direct chitosan-mediated gene delivery to the rabbit knee joints in vitro and in vivo. *Biochem. Biophys. Res. Commun.*, 341, pp. 202-208.

Index